The vintage ROLEX FIELD MANUAL

CHEVALIER EDITION

CREDITS & ACKNOWLEDGEMENTS

Many people have given generously of their time and effort to create this book and I would like to thank Watch Geek-YouTube, Jan Johnson (watchblog.dk), Benn.ch, Jocke, Orchi Pilar, Grenville Lawrence, John Torcasio, Tanner Morehouse, Phillip Ridley, Mike Hui, James Hyman , Philip Corneille (MoonwatchUniverse), Matthew Lobb, Benjamin Coates, Dominic Byrne, Melvin Hollenberg, Flickr The Commons, Wikimedia Commons, Imperial War Museum, The Internet Archive CC, Unsplash Creative Commons, Pixabay CC, Vecteezy.

COPYRIGHT

Library of Congress Control Number: 2020906057

LEGAL NOTE

This guide is an unauthorized work compiled without help from Rolex, associated companies, or serving employees. It is an independent initiative with no legal, economic or personal ties to the famous luxury watchmaker.

It is not sponsored by, associated with, or affiliated with Rolex S.A., Rolex USA or any respective subsidiaries or affiliates.

The following are trademarks of the Rolex Watch Company Limited, Geneva, Switzerland: Rolex, the coronet logo, Oyster, Oysterquartz, Sea-Dweller, Explorer, Submariner, Tridor, Jubilee, Milgauss, Air-King, Cellini, Tudor, Tudor Prince, Tudor Princess, Tudor Day-Date, Mini-Sub, Lady-Sub, Tudor Monarch, and the Tudor shield logo. The author acknowledges the intellectual property rights of the company and its subsidiary companies and distributors.

Revised IV 2020

DEDICATION

To all those with interests and passions.
And N & B, who I hope will one day want to wear these hunks of metal I leave them. Thanks and gratitude to Carla, a constant inspiration and role model. I love you more than you know.

PREFACE

This guide is for collectors of pre-owned and vintage Rolex watches. It is a reference manual and intended to be concise, factual and data-oriented. It was written for the purpose of purchasing watches rather than a celebration of their beauty or history. The book aims to summarize the product lines and references while highlighting the essential nuances important to watch collectors. Many of these subtle details are of little interest to the casual modern Rolex customer, but crucial to the vintage Rolex buyer.

It aims to fill a gap between the glossy coffee-table book and the watchmaker's technical reference. It summarizes the insider knowledge and experience of several collectors and professionals, streamlining the essential education needed to purchase a vintage Rolex. It is not intended to be a comprehensive biography of Hans Wilsdorf or a historical account of his companies, but a focused look at the watches themselves from a collector's viewpoint. If you're looking for the facts and data necessary to make a good purchase, then this is an essential buyer's survival guide.

The book is intended to be practical and focused on pre-owned watches. Much of the beauty found in these watches is the unique way in which they have aged, surviving their past owners and their adventures. Unlike most coffee-table books, this one focuses on the different patterns of wear and signs of age and use.

This philosophy has heavily influenced the author's picture and photo selection. Preference has been given to pictures of watches that are representative of the majority of examples available for purchase in the middle-market. These watches are considered to have warmth and character. You will not find pictures of museum-grade watches, staged and posed in studio lighting. While beautiful and interesting, that type of watch is beyond the reach of the average collector and enthusiast. Instead you will find pictures of real watches, belonging to real collectors, crossing the workbench of real watchmakers. These images are representative of average pieces you will encounter on the open market.

This book came about as a summary of the notes and data I mined and researched over many years. As much as I would like to claim it is entirely without error, I cannot do so. I have made a serious effort to verify and validate the data with service manuals, catalogs, and sales materials. However, much of the information also comes from secondary sources, such as auction catalogs and online watchmaker forums.

This collective knowledge is summarized here to share with others. I hope to equip the reader with the intellectual firepower to take on the unsavory characters that lurk in the shadows of the vintage watch market.

DATA & VALIDATION

Considerable effort was made to verify the completeness and accuracy of the data in this guide. Verification and validation involved modern crowd-sourcing techniques and quality control practices.

Validating reference descriptions involved Amazon Mechanical Turk (MTurk) Human Intelligence Tests (HIT). Mturk is a platform to coordinate individuals performing internet searches. Their task was to confirm specifications and descriptions of Rolex reference numbers.

The human researchers resolved inconclusive searches based on the predominance of results. Application of judgment (by a non-expert, lay-person) can produce imperfect results, much like the imperfect use of reference number conventions by Rolex themselves.

In particular, rare and obscure references were subject to this human discernment. While the number is few, inconsistency errors are likely to be present.

The HIT results were subject to quality control (QC) verification through a separate process. Individuals contracted through Fiverrr.com performed random sampling validation of the Amazon results.

QC verification involved performing the same internet search approach as the Amazon HIT teams. Approximately 150 references were randomly selected for verification (roughly 1 in 10 of the HIT results). The detected error rate was negligible.

The quality of the data compiled here is acceptable for most collectors, however some data is incomplete. If you have data and a reference you know to be correct but missing, please feel free to send it to me. I am happy to accept corrections and additions and would love to hear from any readers of this guide.

CONTENTS

Chevalier | ,SHevə'lir |

noun ***historical***

- a knight or chivalrous man.
- a member of a certain privileged order such as knighthood or of modern French orders such as the Legion of Honor.
- (Chevalier) ***British historical*** the title for pretenders to the British throne.

Chapter 01

INTRODUCTION

Rolex have been making watches in a variety of styles for over a century. Styles and product lines have come and gone with prevailing tastes and fashions. Groups of similarly styled watches belong to a collection, and collections themselves have come and gone as well.

Each watch within a collection has a reference number (commonly cited simply as "reference" or "ref.", followed by the number in question) and the most passionate watch enthusiasts (and serious nerds) will refer to a particular watch model by its reference number. This unhelpful practice is also common when talking about the internal mechanics and the result is a cryptic jargon-soup of numbers intelligible only to initiated insiders.

Known as the movement, watch engines come in a variety of configurations, each with derivations and variants called a caliber ("cal."). These too have their own unique model and serial numbers.

A caliber may have a particular function like a stopwatch (or chronograph), a calendar, or a second time zone. These additional functions are known as complications.

This cryptic shorthand and jargon can be an infuriating barrier to entry for newcomers. These same enthusiasts and collectors are also fond of nicknames and many of them are childishly silly – from Batman to the Hulk, Big Egg to Little Antlers, and countless others. The peculiar habits of the vintage Rolex community are as strange as the members themselves, and it takes patience to understand them and the watches they collect. In doing so, acquiring their madness is inevitable.

Symptoms include compulsive obsessiveness and disconnection from reality, while acute symptoms include bankruptcy and divorce. If, after learning this, you're still interested in acquiring a vintage Rolex watch, this guide will serve as a Rosetta Stone – a universal translator for the coded language and numbers of this strange obsession. With this book you will be able to acquire the watch of your dreams while hopefully avoiding the symptoms of acute watch fixation.

THERE ARE **1,442** REFERENCES *in this book*
PLUS DOZENS OF MOVEMENTS & BRACELETS

632 CLASSICS
249 VINTAGE
209 ANTIQUE
196 PROFESSIONAL
156 DRESS

There are over 1,400 unique references addressed in this guide. The official number and production volume of watches and movements remains a secret and Rolex has never disclosed official records on the matter.

This information vacuum leaves space for controversy and speculation. The gaps in public knowledge have been the subject of considerable academic research and effort. Most information today comes from inference and the extrapolation of information from surviving catalogs and technical service documents. Occasionally, official hints and indicators will emerge from Rolex, often

> "Your time is limited, so don't waste it living someone else's life. Don't be trapped by dogma – which is living with the results of other people's thinking."
>
> *Steve Jobs*

inadvertently. These are poured over and analyzed in forensic detail by the vintage Rolex community.

This general uncertainty leaves much mystery and controversy surrounding specially issued watches, particularly over examples produced for VIPs and royalty, or the short production runs for commercial clients like Comex and Pan American Airways (Pan Am). There are also special versions for armed forces, and limited production releases tested in small foreign markets, as well as prototypes that never made it to market at all. All of these types of watches can be especially hard to validate, yet have intriguing and fascinating implications.

References (or models of watches) and production volumes varied by era as the fortunes of Rolex ebbed and flowed. Manufacturing processes and management practices had to respond to social and economic influences. These included the two World Wars, depressions, technological disruptions, and rapid changes in tastes and fashions. These events make each period distinct and challenging for collectors and buyers.

Collectors group the watches from these periods into three general groups – Antique, Vintage, and Modern Classic and then into subgenres, such as Bubbleback, Military, Arabic, or Transitional. The definition of each term remains controversial and subjective, despite their everyday use. This is a continual source of confusion for inexperienced buyers and an opportunity to lose large sums of money to a seller misusing a term to describe the watch they're offering.

A Quick Business History

Rolex got its start in England, when German businessman Hans Wilsdorf and his brother-in-law, Alfred Davis, founded the watch brand Wilsdorf & Davis in 1905. Neither were formally trained watchsmiths or watchmakers. Their earliest pocket watches bore a "W&D" hallmark inside the caseback and these are extremely rare today, and worth a small fortune.

At this time, the pocket watch dominated the market, and wristlets, as they were known, were dainty jewelry trinkets for young women. The decision to make large, masculine wristlets for men was a bold one. These early watches were to prove their utility on the wrists of Allied officers, in the bloody trenches of the Western Front.

The Rolex name was registered in 1915 as the First World War was erupting.

While there is romantic speculation about why the company name was changed, it was mostly likely a result of anti-German prejudice. Great Britain at the outbreak of WWI was a hostile business environment for any business name as Germanic as Wilsdorf. The new name, Rolex rolled off the tongue, was short and simple to spell (on small parts like dials) and uniformly pronounced in several languages.

Luxury imports to the UK, and in particular precious metals, were subject to heavy wartime taxes and Wilsdorf and Davis decided to move their headquarters to Switzerland. Co-located among Swiss case and movement makers, Rolex went on to distinguish itself with many design innovations, patents, and industry-firsts.

The Founder

Hans Wilsdorf (22 March, 1881 to 06 June, 1960) had an upbringing marked by tragedy and the early death of his parents. In 1903, at just the age of 24, he became the UK importer of Swiss Aegler watch movements. Rolex was later to acquire this long-running and much larger partner.

His career and business was marked by an obsession with detail and the focused and dogged pursuit of engineering perfection.

After the death of his beloved wife he was left bereft and childless, later bequeathing his life's work to the Hans Wilsdorf Foundation, founded in 1944. This philanthropic, not-for-profit foundation continues to own the Rolex assets and oversees operations to this day.

ANTIQUE ROLEX

- **1905** Founding of Wilsdorf & Davis Ltd. in London, England
- **1906** First wristlets for men by Wilsdorf & Davis
- **1908** First registration of the Rolex trademark in Switzerland
- **1910** World's first Official Timing Certificate awarded to Rolex
- **1912** Registration of Rolex trademark in England
- **1914** Kew Observatory awards a Class A certificate to a Rolex wrist chronometer
- **1915** Wilsdorf & Davis becomes the Rolex Watch Co.
- **1925** Rolex coronet first appears on dials
- **1926** Registered, patented, launched the waterproof Oyster case
- **1927** Mercedes Gleitze swims the English Channel wearing an Oyster wristlet
- **1928** Launch of the Rolex Prince and movement classes Prima, Estra Prima and Ultra Prima
- **1931** Launch of the self-winding Oyster Perpetual
- **1935** Sir Malcolm Campbell wears a Super Balance while breaking the world land speed record
- **1937** Launch of the Rolex Precision (refs. 3003 & 3004) and Centergraph (ref. 3642)
- **1945** Launch of the Datejust, Jubilee bracelet and Tudor Watch Co.
- **1947** Completion of the 100,000th Rolex wrist chronometer

Antique Period Timeline

Antique watches are considered to be those produced until the late 1940s and are physically smaller than we've become accustomed to in the 21st century. Pre-war and war-era examples are rare, fragile, and seldom completely original in their parts. Though beautiful to look at, they're quite delicate and can be impractical for daily wear. Styles include pocket, trench, and dress watches for both men and women.

During this era, Rolex experimented with brands and names, targeting different regions and markets throughout the empire and colonies. There is considerable interest in antique Rolex watches and these experimental sub-brands. They are a specialized domain with repairs and servicing becoming more challenging as time goes on.

Surviving antique Rolex watches are amazing feats of early 20th-century design and engineering. They are particularly remarkable considering the historical context and societal norms of the time. The early 20th century was a difficult period in which to prosper. German businesspeople, in particular, would have found this a challenging time and a hostile marketplace.

Today, antique Rolex watches appeal to ever fewer collectors because of characteristics like size and durability since they tend to be so small and fragile. In spite of this, they are historically significant in any condition and remain highly sought after by a dedicated cadre of elite collectors.

Antique Rolex: pocket watch, purse watch and stop watch

ENAMEL DIALS

Most antique Rolex watches feature enamel dials which are brittle and prone to cracking and chipping. Given their age and fragility, it is rare to find enamel dials in pristine condition. Gold pocket watches are more likely to have pristine enamel dials than hard-worn trench watches. While enamel dials can be repaired with modern restoration techniques, the process is risky and usually imperfect. In most cases, restoration involves stabilizing a damaged dial rather then eliminating signs of damage.

Trench watches are some of the earliest wristlets and are often characterized by red 12 o'clock numerals. This was to help orient the watch and aid readability in low light. This was necessary for those transitioning from pocket watches, where 12 o'clock was by convention aligned with the winding crown and the pocket chain. On a wristlet the 12 o'clock position is offset (counter clockwise) to the winding crown by 90 degrees.

BRAND

Admiralty	1914–1923	Jeweler co-branded pocket watch with half hunter movement.
Aqua	1927–1959	Early waterproof Oysters and marketed alongside the Submarine.
Buick	1940s	Thought to be used in the Canadian market only.
Eaton ¼ Century Club	1930–1950s	Eaton's department store employee service award.
Genex	1920–1933	Registered in 1922 and sold to A. Schild in 1933.
Hans Wilsdorf Geneva	1935–1940	Rarely seen and only on Oyster cases.
Hofex	1920s	A movement brand only.
Ingersol	1920s	Complex history as the brand changed ownership several times.
James Walker Ltd, London	1920s	Distinguished English retailer who also sold Rolex and Marconi.
Lonex	1915–1920	An unsuccessful pocket watch brand.
Marconi	1909–1920	A precursor to Tudor.
Neptune	1930s	Early Oysters and assumed to be Canadian market only.
Oyster	1930s–1940s	Used in Commonwealth countries (Canada, Singapore, India).
Panerai	1930s	Rolex supplied movements, dials and cases for Panerai.
Rolco	1927–1930	Appear through Commonwealth countries.
Rolwatco	1922–1926	Evidence suggest a Ladies brand. Usually 9K gold cases.
Sky-Rocket	1930s	Canadian market.
Solar Aqua		Made for Eaton's department stores and Lund & Blockley.
Tudor	1926	Formally adopted in 1946, deprecating all other brands listed here (exceptions being Solar Aqua and ¼ Century Club).
Turtle	1930s–1942	Made for the Zell Brothers, a retailer in Portland, Cregon. May also include Turtle Timer, Turtle Deluxe, Turtle Perpetual, Turtle Lipton, and Turtle Royal.
Unicorn	1919–1933	A common brand that appears to supersede Marconi. Marketed alongside Rolco. A. Schild bought the brand in 1933.
Viceroy	1925–1935	Trench watch brand.
Victory	1945	Commemorative brand celebrating the WW2 Allied Victory. Sold in Commonwealth countries.
Wilson & Sharp Edinburgh	1920s	A distinguished Scottish retailer and jeweler.
Wintex	1920s	A movement brand only.

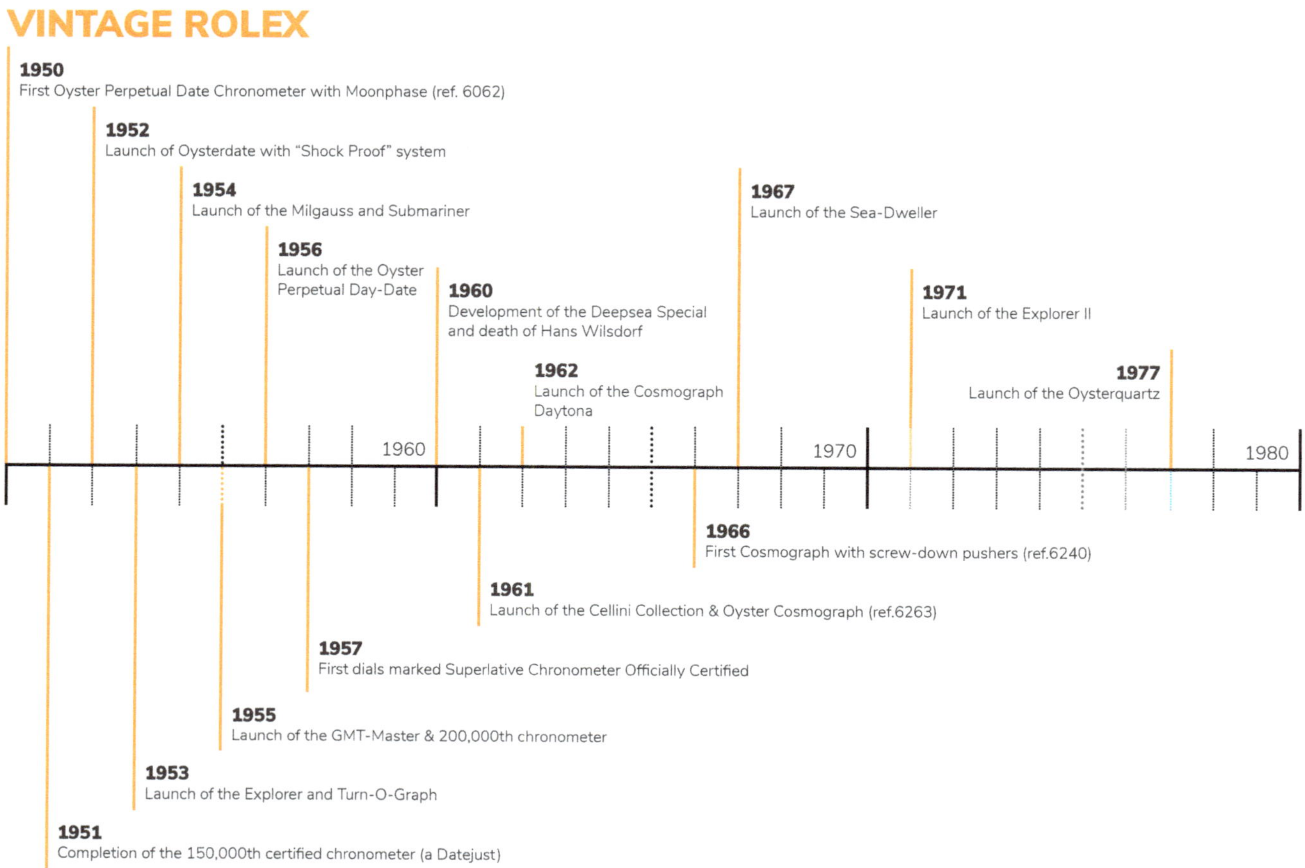

Vintage Period Timeline

Watches made from the early 1950s through the 1970s fall into the vintage era. This period was a particularly prolific time for Rolex. They found commercial success in their new sporting and professional watches. It was also a calm period in world history, allowing Rolex to grow into new markets with new products.

In the absence of war, great men and women were free to pursue remarkable human achievements. The 1950s and 60s were a productive period for these fantastic firsts. The ability to travel freely across borders and having relative prosperity fueled these pursuits and international records were broken all around the world.

Precision timekeeping was crucial in new and modern pursuits like mountaineering, aviation, exploration, space travel, and saturation diving. A wristwatch was a life-critical instrument in extreme environments, and Rolex was on a mission to make the best.

Collectors consider this the Golden Age, as references are still robust and wearable today. Many are historically significant and survive in excellent condition. Examples have travelled from the deepest oceans to the moon and back, on the wrists of aquanauts and astronauts. From mountaineering to motorsport, professional tool watches with documented provenance are highly sought after.

The market for vintage Rolex watches is large and liquid, and prices have been on a steady upward trajectory for several years, attracting new collectors, investors, and speculators. Watches designed as tools for these challenging pursuits lead this price surge. The most popular are those made in stainless steel.

THE HEINIGERS

Andre Heiniger was the second CEO to take the helm of Rolex in 1963, three years after the death of Hans Wilsdorf. He began working at Rolex in 1952, directing operations in Argentina, where his son and successor, Patrick, was born in 1950.

The family moved to Switzerland in the early 1960s before assuming control of the company. He was a staunch traditionalist, only reluctantly embracing the Oysterquartz in the face of mounting pressure to respond to the industry quartz-crisis.

His son, Patrick, became Commercial Director of Rolex in 1986, and after the death of his father in 1992, took over as Managing Director. He went on to become the CEO in 1997.

Patrick was responsible for the aggressive industrialization and vertical supply chain integration that enabled international expansion.

MODERN CLASSIC ROLEX

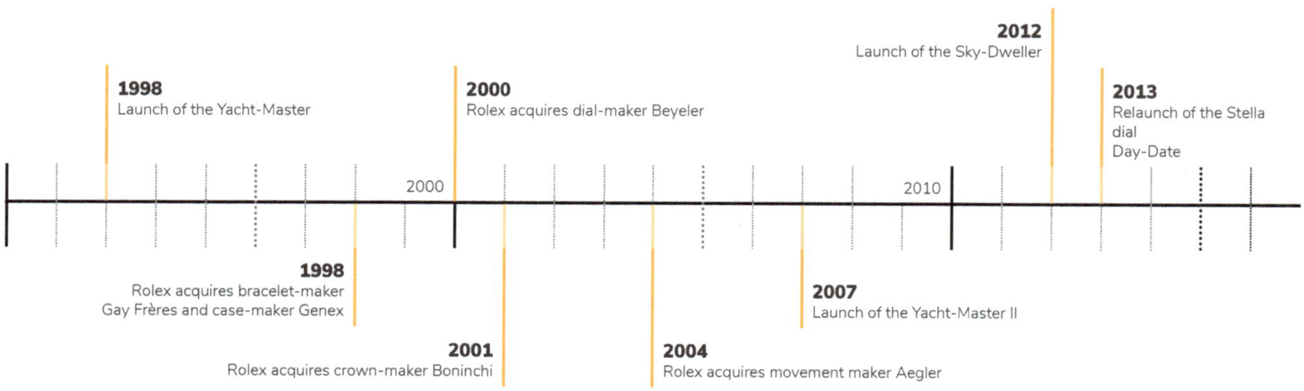

Modern Classic Period Timeline

Modern Classic watches are those from the 1980s to the turn of the century. The term is unofficial and coined by collectors and shouldn't be confused with the Rolex Classic Collection or classically styled models. This genre of watches is generally an evolution of iconic vintage designs. Rolex service centers can still work on them, thanks to their contemporary movements and parts availability. Serviceability makes them accessible and practical for new collectors.

This category of modern classic watches has won the hearts of new, younger collectors. These less seasoned enthusiasts often lack the experience, the means, and the knowledge to dive straight into more technical vintage examples. The audience for Modern Classics is the fastest growing Rolex market. It is liquid and still offers reasonable price and value today.

This era was a period of aggressive business acquisitions by the third Rolex CEO, Patrick Heiniger. His goal to achieve vertical integration of the supply chain was mostly successful. The rate at which he was able to acquire and integrate subcontractors, suppliers, and partners is a masterful leadership and management achievement. The result is a more consistent product with fewer defects and anomalies.

The variations and defects that characterized the vintage period are much less common in Modern Classic watches thanks to streamlined production and improved quality controls. Subsequently, watches of this period are more uniform in quality and consistency.

CONTEMPORARY ROLEX COLLECTIONS

Oyster Perpetual Classic	Oyster Perpetual Professional	Cellini
Oyster Perpetual	Cosmograph Daytona	Cellini Time
Datejust	Sea-Dweller	Cellini Date
Day-Date	Submariner	Cellini Dual Time
Pearlmaster	GMT-Master II	Cellini Moonphase
Sky-Dweller	Explorer	
	Explorer II	
	Milgauss	
	Yacht-Master	
	Yacht-Master II	

Contemporary Rolex Collections

Historical significance is an attractive characteristic of antique, vintage, and classic Rolex watches. Amazing human achievement enabled by precision timekeeping adds to the general aura and mystique. Calculating flight time, oxygen, daylight, and other life-critical functions burnish their reputations. Horological history recognizes them as essential tools and their modern descendants proudly claim them as part of their genealogy.

The current family tree of products gets much of their value and mystique from these old watches. Each current Collection has examples of these ancestors with numerous iterations and descendants. Distinct model reference numbers are used to identify each of them.

Reference numbering conventions have evolved. Even in stable periods of production, there are strange inconsistencies. These anomalies make collecting challenging and uniquely entertaining.

Although the focus of this guide is pre-owned antique and vintage Rolex watches, context and completeness requires the inclusion of some contemporary references from the current Collections.

THE OYSTER CASE & CROWN

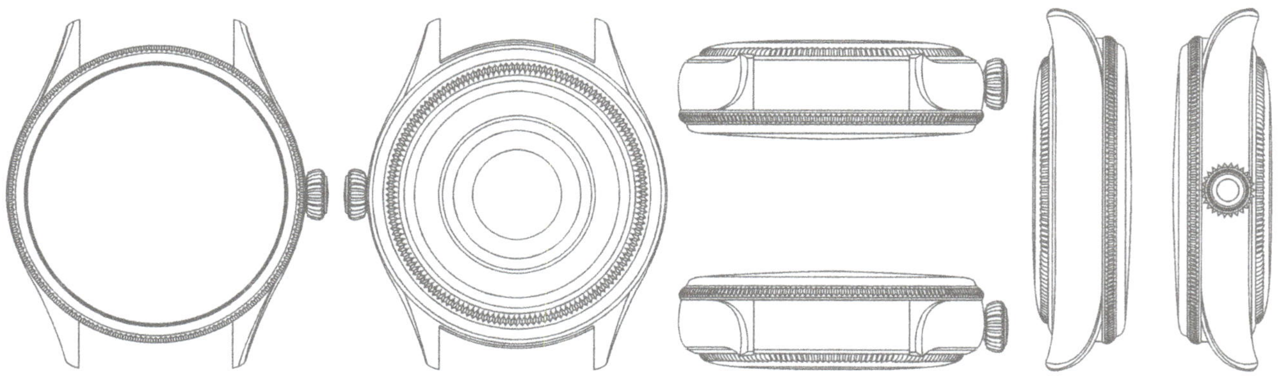

US Patent D733,582S for the Rolex Oyster Case

The name Oyster is most commonly associated with the iconic waterproof case. Few know that it was also used to brand a collection of early watches in 1926. Later sections of this guide discuss this initial Rolex Oyster collection.

Waterproof watchcases were in use well before Rolex introduced the Oyster. Dennison (a British watch case maker) secured a UK patent in 1872. Ezra Fitch (the co-founder of the modern lifestyle brand Abercrombie & Fitch) was awarded a US patent in 1879 and 1881.

These patents and designs were combined and used in the Alcide Droz & Fils Impermeable, a waterproof pocket watch made in 1883.

In 1917 two Royal Navy submarine commanders commissioned the Submarine Commanders Watch. It was a watertight wristwatch with a screw-on case back and bezel, both with compressible gaskets. It also had a waterproof compressible seal in the stem tube to avoid water ingress from the crown. While the configuration is now familiar to Rolex, it was not a commercial success in its day due to poor marketing and business practices.

Alcide Droz & Fils Impermeable

In October 1925 Paul Perregaux and Georges Perret registered Swiss patent No. 114948 for a crown that could be screwed onto the case to create a waterproof seal. It was this patent that Hans Wilsdorf acquired the rights to and subsequently extended and improved. Less than a year later in 1926 Wilsdorf registered patent No. 120848 and paired the design with his Oyster case.

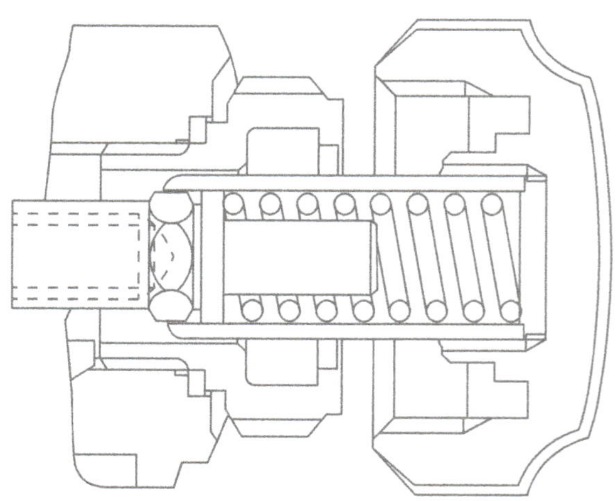

Sectional view from patent drawing of Triplock crown

THE PERPETUAL MOVEMENT

The story of the Oyster case is intimately related to the invention of the automatic self-winding Perpetual movement. In times past, it was common for owners to forget to screw down the crown and hermetically seal the case. Forgetfulness would result in terminal damage to the inner workings of the watch. Rolex concluded that the solution was to eliminate the need to unscrew the crown and wind the watch in the first place. This lead to the aggressive pursuit of an automatic self-winding movement.

Rolex first used the term Perpetual with their early self-winding bumper movements in 1931. A bumper movement uses a spring-loaded weight that oscillates back and forth (or up and down), as opposed to a rotary weight that swings around a central axis. The rotary design was by no means an industry first, as that credit goes to John Harwood, of the Isle of Man in 1923. He was unable to commercialize his invention or exercise the patent, leaving the field open to players like Rolex.

By 1931 Rolex had laid the foundations for success by acquiring the necessary patents and making their own small and incremental design improvements. This lead to the 360-degree winding rotor retrofitted to a conventional manual movement, enclosed in a water and dustproof case. The height and thickness of the resulting package were heftier than competing products, many of which had more technically sophisticated designs.

This oversized hack of a product should not have been a market-beating item because a conventional manual wound movement with a retrofitted self-winding mechanism resulted in a large, bulky package. However, with marketing prowess and an uncanny ability to influence customer perception, Rolex was set to dominate the industry. They offered something robust and substantial on the wrist that didn't need winding (but could endure one anyway) and was impervious to the elements. It wasn't haute horology, but it was a practical and effective compromise, packaged cleverly and launched quickly. And so nearly a century later, the words "Oyster" and "Perpetual" adorn almost every Rolex dial that leaves the factory.

Movement (left), auto-wind assembly (right top) and semi-circular rotor weight (bottom right)

Vintage | vïn'tïj |

adjective ***historical***

- characterized by excellence, maturity and enduring appeal.
- a group or collection of people or things sharing certain qualities.
- output of a particular time now considered old or outmoded (retro).
- being of the best of its kind.

Chapter

02

AUTHENTIC, ORIGINAL & CORRECT

The vintage Rolex watch community is diverse, and the range of views they hold is broad and often contradictory. There exists (particularly on the internet) an unhealthy and exaggerated obsession with originality, authenticity, and correctness. While these things matter, newcomers should maintain some perspective by looking for context and remembering their priorities.

Most of us are seeking 1) an authentic Rolex, 2) in a cosmetically attractive condition, and 3) in good running order. While this guide won't help you to identify a fake, it will reduce your odds of being conned while you try to meet these criteria and stay within budget.

Fake or replica watches are assembled with the intent to deceive. Refinishing and restoration are sometimes done in the same spirit – to pass the piece off as entirely original. While counterfeiting is illegal, undisclosed restorations are not, and we should be looking for signs of undisclosed enhancement or modification.

"The best and most beautiful things in the world cannot be seen or even touched – they must be felt with the heart."

Helen Keller

AUTHENTICITY

Many authoritative sources can authenticate a vintage Rolex. The best places to go for authentication are authorized Rolex Service Centers (RSCs) and acknowledged independent vintage Rolex specialists. Authorized dealers (ADs) and general retailers are not skilled at authenticating antique and vintage Rolex watches.

Authenticity describes whether a watch contains any non-Rolex, counterfeit or aftermarket parts. Authenticity is a question of degree and can range from outright counterfeit at one end of the spectrum, to a genuine Rolex with a few after-market parts on the other. The term should be used carefully with the understanding that it is not always clear-cut.

Authentication requires opening the caseback and examining the mechanical movement. An experienced watchmaker should perform this in person with the appropriate tools.

If this isn't possible, other options exist. Begin by taking well lit and focused photos from all angles. Remove the strap or bracelet and get pictures of the engravings between the lugs. If possible, open the caseback and get pictures of the markings and the movement too, then submit these photos to vintagerolexforums.com or rolexforums.com. Experienced collectors and enthusiasts will share their often brutally honest opinions. A range of views can help a collector decide the degree of authenticity.

Rolex Service Center, Prague, Czech Republic

ORIGINALITY

When talking about originality, we are assuming everything is authentic. The question of originality concerns the presence of service or donor parts, and while they may be genuine, they're installed after the watch leaves the factory.

Originality matters most to a small subset of elite collectors. These folks are interested in the top end of the market for rare, high-value examples – particularly those in unused, mint, and otherwise unmodified condition. These watches will have spent decades locked away in a safe, and remain in New Old Stock (NOS) condition. NOS is a brand new, unsold watch from old inventory.

The overwhelming majority of vintage watches available to average enthusiasts are not NOS. They have been used and worn, acquiring a charm and character unique to each piece. For these watches, the issue of originality matters much less than other factors like condition, correctness, and authenticity.

The term All Original means that a watch contains only the parts it left the factory with. This label implies it has never been opened or serviced. These are museum grade examples and not something you'd want to wear.

Modern restoration and refinishing services can return a watch to almost any condition desired, including NOS. While this work is mostly honest, legitimate restoration, it can be also be performed in an underhanded and dishonest way. If the work is undisclosed and intended to deceive, it is the equivalent of turning back a car's odometer. With refinishing practically impossible to detect, many claims of being All Original are usually either misguided or dishonest. Originality can only be assured with any degree of certainty through forensic analysis in partnership with Rolex.

CORRECTNESS

Correctness describes whether all the authentic Rolex parts are the right ones for the reference and belong together. This is referred to as period correctness.

It's common to find caseback reference numbers that don't match the mid-case reference. Many casebacks are interchangeable and can get unintentionally mixed up while on a watchmaker's bench. Sometimes parts aren't available, and the watchmaker has to use whatever fits.

It is also common to see a bezel insert that is not appropriate for the age of the watch or not originally offered on a particular reference. An incorrect bezel is common on watches like the GMT-Master, where owners have changed colored bezel inserts out of personal preference.

You might also see a watch paired with a bracelet that was never an official option. While switched-out Rolex parts shouldn't be a deal-breaker for a watch you like, recognizing them can help to understand the value of the piece and negotiate a fair price.

TRANSITIONS

Swiss watchmakers, in general, have a frugal and cost-conscious culture. Parts and components were often designed to be shared and reused across model lines and references. Introduction of a new reference often involves using up surplus inventory and many watch manufacturers engage in this frugal practice.

These curious watches with their anomalous configurations are known as Transitionals. Transitionals are those models that include features and parts of previous or future references. Collectors are divided over whether they are more or less desirable or collectible. While they are made in relatively small volumes and are often rare, they don't command any price premium over non-transitional models despite being rarer.

It's common to see a caseback on a Transitional with a reference number crossed out and a new later one applied. This correction is of no concern if the serial number dates the watch to a known transitional timeframe. You can also expect to see old dials and movements paired with newer cases (e.g. Submariner ref. 168000), or new movements deployed into outgoing references (e.g. GMT-Master ref. 16710 with cal. 3135). Inexperienced collectors will often brush these off as fake, or "Frankenwatches", assembled from surplus or leftover parts. Informed collectors will recognize them as an opportunity to acquire a rare and oftentimes underpriced, authentic and correct watch.

All mature Rolex product lines have acknowledged Transitional examples, but few have distinct reference numbers, like the Submariner ref. 168000, the only Transitional with its own unique reference number. Unfortunately, the reason that this specific Transitional has its own number is unknown. Transitionals are not always easy to identify, and legitimate transitional models can be challenging to assess for originality. However, you will find data in later chapters that will help evaluate the authenticity and correctness of Transitional references.

Transitional Submariner Date ref. 168000

Crossed out caseback

PERSPECTIVE

Confirming any pre-owned Rolex watch as definitively authentic, all original, and entirely correct is virtually impossible without the aid of advanced forensic lab equipment (like radio spectrometers) and an abundance of verified historical data. You must accept that a level of risk is present and that even elite collectors struggle to reach certainty. You will especially need to get comfortable with this idea if you wish to avoid analysis paralysis or buyer's remorse.

The overwhelming majority of vintage watches have experienced maintenance servicing during their lifetimes. With Rolex service parts dwindling or now extinct, aftermarket and even handmade parts are often used to keep old watches running. Genuine Rolex service parts installed during maintenance are acceptable, but less desirable than original factory-installed parts, and softer prices for watches with service parts reflect this preference.

If a watch with service parts is priced lower than one with original parts, consider this...

There is a thriving global market for watch components that have been parted-out. This term means a watch is broken down, and its parts sold individually.

If a vintage watch has some of these donor parts installed rather than identifiable new service parts, is it still original? Would you care? Would anyone be able to prove parts come from another Rolex watch? Do you personally prefer newer components that are genuine service replacements or older looking, well worn, donor parts? Knowing there might be mismatched donor or service components, would you pay a premium for a well-used watch that claims to be all original?

Like vintage cars, collectors accept that non-original parts are required to keep an old car running and valuable. Aftermarket parts might include tires, filters, hoses, and even paint. Comparable consumable parts for a vintage watch would be barrel springs, crystals, crowns, tubes, gaskets, bracelets, clasps, and a host of moving parts. A vintage watch may no longer be all original, but it can still be beautiful, functional, and highly desirable.

There are many legitimate, minute variations within references. These include crown size, crown markings, dial text, hour markers, bezels, inserts, and hands to name a few. These variations extend to customizations like co-branded dials with favored jewelers and retail partners (such as Serpico y Laino, Joyería Riviera, and Tiffany).

There are further variations in Rolex service parts, which have also evolved over the years. All these variables complicate any assessment of authenticity, originality, or correctness. Newcomers to vintage Rolex collecting are encouraged to embrace the ambiguity and mystery as part of the unique character of each vintage watch. These signs of a well-owned vintage Rolex should be welcomed rather than shunned.

Rolex Submariner with Tiffany Dial

A class of watch known as marriage watches also exists. These are Rolex pocket watch movements re-cased in wristwatch cases. While not original, they are legitimate and can be beautiful and artistically executed. New collectors are advised to stay away from marriage watches, as they're a very specialized niche domain.

Cosmographe | ˌkos'mo.gra.f |

noun ***scientific,*** Ancient Greek, Latin

- one who describes the universe.
- a representation of the world or universe.
- the science dealing with the whole order of nature.

Chapter 03 ACQUISITION

If you're interested in vintage Rolex watches, you will eventually ask, "How much are they?" It won't be long before you learn how technical the question is and see that much depends on whether the intent is to buy or sell and if the counter-parties are businesses or individuals. Either way, valuations and opinions vary widely.

Prospective buyers should very carefully consider the difference between price and value, and what each means to them personally. They can begin by examining their own motives. Are they looking to mark an occasion with a piece of jewelry, or find a stable cache of wealth? Are they planning to build a long-term collection, or find a gift for someone special? Is this their first Rolex, their last, or something in between?

These questions shape how we think of value and price. Neglecting introspection and merely riding the wave of shopper's enthusiasm will result in unexpected surprises and disappointments.

If a price is the money necessary to acquire a watch, then the value is the desirability of that watch in terms of its exchangeability. While they are related, they should not be confused with one another.

For example, a stainless steel Submariner ref. 1680/1 will command a lower price than its solid gold counterpart, the ref. 1680/8 – but the stainless steel version is considered more valuable due to higher demand and the ease with which it can be resold. The value, in this case, is based on market size and liquidity, rather than the raw material of which it's composed.

While the solid gold 1680/8 is beautiful and desirable, fewer collectors can afford them, and they can take much longer to resell. Prevailing tastes and attitudes towards a heavy, solid gold Rolex also vary widely by region, therefore, the market for the 1680/8 is smaller and less liquid.

For a prospective buyer, value involves the peace of mind, communication, service, and convenience. Value to a seller includes transaction speed, low fees, and commissions, repeat or volume business, and reputational gain.

For both buyer and seller, there are risks to be calculated. They must each take into account what they know about the watch itself and about each other. This requires patient research and delving into historical sales data. While many will tell you to just "buy the seller" (meaning a seller's reputation), this is only one factor in quite a complicated equation. Sources of price data include auction house sales, eBay, Chrono24 and watch forum listings.

Sources of value data include reputation from feedback, endorsements, and recommendations. These might be found on Yelp and BBB ratings for business sellers. Professional membership and accreditation like AWCI and NAWCC are seller credentials that can also be verified. AWCI (American Watchmakers-Clockmakers Institute) exists to set service standards and educate the horological community. NAWCC (National Association of Watch & Clock Collectors) exists to educate and promote ethical practices. General market conditions can be assessed by conversing with dealers and collectors at meetups and online forums.

Even with careful research, the risk will never be zero. If a watch is well bought, long-term risks inherent to vintage watch ownership still remain. While vintage Rolex prices hold up better than competing brands, this is no guarantee of sustained long-term price appreciation. You must factor in the rising cost of servicing and the declining availability of parts, plus economic conditions and current tastes. It is definitely possible to lose money on a well-bought vintage Rolex, and there are no rock-solid guarantees of any type in vintage Rolex ownership.

"The road to success and the road to failure are almost exactly the same."
Colin R. Davis

MARKETS

The location of a watch you want to buy can complicate your price and value calculation. If the piece is crossing national borders, either on your wrist or via a courier service, there are costs and risks to consider.

It is important to recognize that regional markets (e.g. EU, the Americas, Asia, Far East) have seasonal and local tastes and customs which vary widely. For example, prices and exchange rates tend to surge during holiday and bonus seasons. This boom runs from around Thanksgiving in November through Chinese New Year in February. Trends vary too; steel sports references like Submariners, GMTs and Explorers are more popular in North America and Europe. Asian and Eastern markets often favor smaller cases forged from precious metals, in part due to the cultural significance of gold, the way darker skin tones look with different-colored metals and that slender wrists are more common. Having said this, Asian collectors are among the most informed, aggressive and ambitious – particularly at the elite and top end of the market.

For the price-sensitive collector in the middle and lower end of the market, sales taxes, import duty, shipping and payment processing fees can provide further price variables. For example, the UK VAT (Value Added Sales Tax) at 20% is required to be included in the dealers' asking price. In the US, sales tax is applied separately to the asking price. The UK import duty on a vintage watch is not the same as the US import duty on that same watch. Local shipping rates and courier insurance also vary, with shipping from the UK to the US via UK Royal Mail being less expensive than shipping from the US to the UK via USPS. These fees can be significant and unexpected additions to the total cost of buying a watch.

Recourse and protections for a watch purchased from overseas are, in practice, close to none. Shipper's insurance is tricky to claim against if your purchase is lost or damaged in transit. All the major courier companies specify the maximum they will pay out on an insurance claim in their terms and conditions (the small print). Buying additional insurance is seldom worth it.

While eBay and PayPal offer some buyer protection, exercising these rights is challenging and will require that you submit comprehensive documentation and communication records. Credit card protections on overseas internet purchases are also challenging to use, and procedures and coverage vary depending on the card issuing company.

Even if a site or seller claims to have an abundance of happy international clients and global endorsements, you should know there is a risk, and you should have a plan in case things go wrong. This should include retaining copies of all communications, terms and conditions and any photos used to sell the piece. Select a method of payment carefully and understand any insurance coverage and dispute or arbitration process that may be available (credit cards are safer than wire transfers). If you're shipping a watch you've sold, photograph the piece at every stage of wrapping to dispatch and don't ship before funds have cleared (or escrowed).

WHERE TO PURCHASE

The balance of risk is always higher for a buyer. The best purchases are achieved by maximizing trust and transacting in person. This is known euphemistically as the friends & family trade.

The next best type of purchase is with a gray dealer with whom you have transacted before and developed trust and rapport, preferably in person. Gray dealers are independent, unofficial and unaffiliated with Rolex. They acquire their inventory from Rolex Authorized Dealers at discounted trade prices through trade channels like watch fairs. They are typically known for selling new watches, but often have access to, or hold vintage pieces.

Many gray dealers use online forums as sales channels and invest heavily in their online reputation. The value of a gray dealer is the diverse inventory they provide to the buyer through their access to their wider network. Concierge services can also be especially valuable. This service involves seeking out a watch in a specific condition and price, for a retainer or engagement fee.

The next best point of purchase is through a vintage retailer such as a long established brick and mortar business or the popular online stores. Popular and reputable sites include HQMilton and BobsWatches. Buyers rely heavily on their descriptions, photos, reputations, and return policies. Expect to be low-balled on pieces you are trying to sell and charged what they think the market will bear on inventory they offer. The spread can be wide depending on the model (typically 8% to 20%).

The last is the Rolex authorized dealer. Few are interested in selling pre-owned or vintage pieces, preferring to unload this inventory through trade channels to the pre-owned dealer network. These outlets may occasionally (and unofficially) have pre-owned watches they've received in partial exchange on new purchases. They generally have very little expertise in the highly technical vintage market. The retail sales personnel are incented to sell new Rolex watches to maintain their employer's authorized dealer status.

RESEARCH & DUE DILIGENCE TIPS

An excellent way to determine the prevailing price of a specific vintage watch is to trawl the internet collecting prices from wherever they can be found. This can include watches currently offered for sale as well as sold items. Seasoned collectors tend to enjoy this education and research while less experienced buyers can find it frustratingly tedious and confusing. In other words, the learning curve is steep but eventually gets easier.

As you collect this price data, try to account for the presence and condition of bracelets, boxes, papers, hang tags, booklets, and other accessories.

It would be sensible to limit the time period of items for sale to say, the last 12 months, or even less for really desirable references. This will enable you to calculate an average price for an average condition in the current market – a useful benchmark. eBay has a particularly useful search filter that allows you to search sold items and their final selling price.

The next step is to seek a third party opinion on the condition, originality, and desirability of a particular watch. You should not ask, "What is this watch worth?" but instead ask, "How do you rate the condition of this watch?" Opinions on condition will help you place it above or below the average benchmark you calculated earlier. Expect opinions to be harsh and widely varied. Remember, you are seeking a general consensus of views, or what most observers think of that particular watch – is it better or worse than average? Is it above or below your benchmark?

It is also worth consulting others about local market conditions and local interest in a particular reference. If considering an international transaction, seek out the experience of others, particularly those with logistics and duty expertise.

If you don't move in watch collector circles, make use of the two most crucial online watch forums for Rolex collectors. These are considered authoritative and are frequented by seasoned collectors as well as watchmakers specializing in vintage references. While the members' opinions and expertise are invaluable, the forum format can make research frustrating and tedious.

- The Rolex Forums
  rolexforums.com
- Vintage Rolex Forums
  vintagerolexforums.com

POST-PURCHASE

Once you've acquired a vintage Rolex watch, your first duty as the new owner is to ensure any damage is stabilized and the watch remains in reliable running condition for years to come. This will invariably involve a visit to a watchmaker for various degrees of work ranging from a light service of lubrication and calibration to comprehensive restoration.

A MODERN WATCH FLIPPER

Tanner Morehouse, North Dakota, USA

Buying luxury vintage watches from unconventional sources offers the opportunity to pick up incredible bargains. The most notable of these are online at auctions like Ebay, Gumtree, Facebook, Craigslist and others.

But beware, you are likely up against unscrupulous sellers and stiff competition. Not just deep-pocketed enthusiasts but highly skilled and knowledgeable professionals. One such example is Tanner Morehouse, a young, professional and upwardly mobile watchmaker.

Tanner's main line of business is full-spectrum watch servicing, specializing in vintage Omega and Rolex restorations. Tanner is a self-taught watchmaker who has acquired not only a high level of technical skill but also deep expertise, earning him a loyal and committed clientele.

As a sideline, Tanner has become a master at buying distressed watches from some of the most cutthroat auctions, to restore at his leisure at his own cutting edge facility. He has years of hard-earned experience and a few expensive mistakes under his belt, making him uniquely qualified to help you acquire or assess a vintage piece you may have your eye on. Buyers often overnight ship their newly purchased vintage watch to him for assessment and servicing. If they've bought a lemon, they can count on Tanner to quickly and tactfully tell them so!

"It's easy to find a beautiful and potentially rare vintage watch. It is much harder to acquire it at a price that's fair and reflects the level of risk. If you've only got photos to go on, you need the experience of having handled these watches under magnification to interpret what you might or might not see in a poor photo. If you're planning on spending serious money, it helps to have a network of friends and professionals you can turn to quickly for an honest second opinion before the gavel drops!"

Unlike other young vintage watch influencers, Tanner maintains a relatively low social and professional profile. His down to earth and straight-talking style has endeared him to vintage collectors, who find his enthusiasm infectious and his knowledge impressive. If you're looking to purchase a vintage Omega or Rolex, Tanner can be reached online at tmwatchco.com

Chapter 04

BY THE NUMBERS

Vintage Rolex model reference numbers are engraved rather than stamped on the Oyster case between the lugs on the 12 o'clock side. Serial numbers are on the 6 o'clock side. This practice is limited to Oyster cases. Non-Oyster cases such as Cellini and antique Rolex, like early Bubblebacks, have their reference and serial numbers engraved on their casebacks.

On Oyster cases, there is considerable variation in the exact positioning and font of these engravings. They were written using a mechanical Pantograph, and the text is not as consistent and uniform as that of computerized laser etching. Furthermore, some previous Rolex Service Centers would re-engrave numbers if they were worn off or becoming too hard to read. Identifying authentic and original engravings requires experience and practice, and even then the results can be debatable.

Engravings can be obscured by dirt and corrosion or worn away by poorly fitted end links (often the result of incorrectly sized spring bars). In cases such as these, a jeweler's loupe is required to read them. Collectors should be on the lookout for signs of re-engraving, over-engraving, or other attempts to tamper with these markings. Any corruption of these engravings will seriously devalue a vintage Rolex.

The conventions used for these reference numbers are inconsistent. While generalizations are possible, many exceptions exist. Some models are dual-referenced, meaning the same case reference number can be assigned to two different models. For example, some vintage Air-Kings carry the same case reference number as vintage Explorers. This reference number can usually be split into three parts, indicating the model line, the bezel style, and the case material. For example a GMT with ref. 16710 would be 167 denoting a GMT-Master II, 1 indicating a steel bezel (not the bezel insert) and 0 indicating a stainless steel case.

> "A good decision is based on knowledge and not on numbers."
>
> *Plato*

Meta engravings

When examining reference numbers between lugs, it is common to see additional text of the form, "REGISTERED DESIGN" or "ORIG ROLEX DESIGN".

"REGISTERED DESIGN" appears on watches up until 1980 or the 6.2M serial range. From approximately 1980–1982 (6.2M to 7.0M) the text was dropped with just the model number engraved. This marked the expiration of the Oyster patent. Then from 1982 (7M) onwards, the words "ORIG ROLEX DESIGN" were used.

An authentic vintage Submariner that appears to have been subject to re-engraving

Engravings obscured by poorly fitting end links

The brushed finish (vertical stripes) are a distinguishing characteristic of the way Oyster cases are made and can be challenging to fake

CASE MATERIAL KEY IDENTIFIERS

Occasionally you'll see old reference numbers with a suffix separated by a slash ("/") – for example, a ref. 1675/8 with the /8 denoting an 18K solid gold case. The use of the slash is not consistent but appears only on four-digit references.

By the time Modern Classics emerged with five digit reference numbers, the slash had fallen out of use. So our example would become a ref. 16758.

The following table will help translate that last digit and can be used to validate warranty papers and confirm they match the accompanying watch.

Reference Suffix	Case Material
0	Stainless Steel
1	Yellow Gold Filled
2	White Gold Filled
22	Stainless Steel & Platinum
3	Stainless Steel & Yellow Gold
4	Stainless Steel & 18K White Gold
5	Gold Shell or Everose (Rose Gold)
6	Platinum
7	14K Yellow Gold
8	18K Yellow Gold
9	18K White Gold

MODEL COLLECTION IDENTIFIERS

A reference number prefix can be used to validate which Collection a watch belongs to. This is relevant because Oyster cases come in standardized sizes and can be reused across models. If you're looking for an original watch, the case prefix should reflect the model on the dial. For example, if you are looking at a vintage Submariner with a reference number of 167x, then you are looking at a Frankenwatch that has been assembled from a parts bin.

Model	Reference Prefix
Submariner No Date	55, 140
Submariner Date	16, 166, 168
Sea-Dweller	16, 166
GMT-Master	16, 65, 167
GMT-Master II	167, 1167
Day-Date	65, 66, 18, 180, 182, 183
Datejust	16, 162
Daytona (Manual)	62
Daytona Cosmograph	165, 1165
Explorer II	165
Oyster Perpetual	10, 140, 142
Air-King	55, 140
Date	15, 150
Oysterquartz Datejust	170
Oysterquartz Day-Date	190
Yacht-Master	166, 686, 696
Datejust (Mid Size)	68, 682
Oyster Perpetual (Ladies Size)	67, 671, 672
Date (Ladies Size)	65, 69, 691, 692
Datejust (Ladies Size)	65, 69, 691, 692
Oysterdate (Mid Size)	65

MODEL COLOR CODES

Occasionally you'll see a model reference with letters at the end. This convention is applied mainly to modern models and appears on warranty papers only (not on the watch itself). Rolex and their authorized dealers also use them for administrative purposes.

Sometimes you'll also see these abbreviations used as informal shorthand and applied retrospectively to Modern Classics. Most commonly you'll see this in for sale ads.

Reference Suffix Code	Color Code Translation
BLRO	Bleu-Rouge = Blue & Red Bezel (Pepsi)
BLNR	Bleu-Noir - Blue & Black Bezel (Coke)
LN	Lunette Noire = Black Bezel
LV	Lunette Verte = Green Bezel
LB	Lunette Bleu = Blue Bezel
GV	Glace Verte = Green Sapphire Crystal (Milgauss)
RNBW	Rainbow = Multicolored Sapphire Bezel

CASE SERIAL NUMBERS

Each Rolex watch leaves the factory with a unique serial number that can be used to approximate the production date. However, the numbering format can be inconsistent, and there are exceptions, with some antique watches having no serial number at all.

In 1954, Rolex reset its serial numbers after they reached 999,999 making it possible for a watch made in the 1960s to have the same serial number as one from the 1950s. To help distinguish between the duplicates, Rolex engraved case backs with date codes, making it easier to differentiate between a pre-1954 and a post-1954 serial number. However, it's possible that casebacks could get mixed up on a watchmaker's bench, causing problems dating the watch and assigning a value.

When serials reached 9,999,999 by mid-1987, Rolex began using a character prefix followed by six digits. Curiously, they did not start with the letter A. Instead, they started with R, followed by L, then E, followed by other non-sequential characters.

The most significant change in serial number conventions came at the end of 2010 when Rolex adopted random serial numbers, making it impossible to determine a year of production.

The prevailing theory is that Rolex did not want to disclose annual production volume to competitors or customers, while others speculate it was an anti-counterfeiting measure.

A watch's serial number can be found on its warranty papers as well as on the watch itself. Watches made before 2007 will have the serial number engraved on the case in between the lugs at 6 o'clock. Seeing the number requires removing the bracelet, and depending on the case's condition, a jeweler's loupe may be needed to read it.

In late 2006, Rolex began laser etching the serial number on the rehaut (the perpendicular ring around the outer edge of the dial). In 2008, Rolex phased out the case engraving of serial numbers and began relying exclusively on the rehaut etching.

The serial number on the case is unrelated to the serial number on the movement. The mechanical movements were manufactured under different processes, and in some cases by different companies (e.g. Aigler, Valjoux). Production and assembly records, which match case serials with movement serials, remain a closely guarded secret. These records are available only to Rolex employees at Rolex Service Centers. There is an ongoing debate about how complete and accurate they are for vintage watches.

2010s		2000s		1990s		1980s		1970s	
2019	Random	2009	V	1999	A,000,001	1989	L,980,000	1979	5,737,030
2018	Random	2008	M OR V	1998	U,932,144	1988	R,598,200	1978	5,000,000
2017	Random	2007	M OR Z	1997	U,000,001	1987	R,000,001	1977	5,008,000
2016	Random	2006	D OR Z	1996	T,000,001	1987	9,400,000	1976	4,115,299
2015	Random	2005	D	1995	W,000,001	1986	8,900,000	1975	3,862,196
2014	Random	2005	F	1994	S,860,880	1985	8,614,000	1974	3,567,927
2013	Random	2004	F	1993	S,000,001	1984	8,070,022	1973	3,200,268
2012	Random	2003	F	1992	C,000,001	1983	7,400,000	1972	2,890,459
2011	Random	2002	Y	1991	N,000,001	1982	7,100,000	1971	2,589,295
2010	G	2001	K OR Y	1991	X,000,001	1981	6,520,870	1970	2,241,882
		2000	K,000,001	1990	E,000,001	1980	6,434,000		
		2000	P,000,001						

1960s		1950s		1940s		1920s		1920s	
1969	1,900,000	1959	399,453	1949		1939	71,224	1929	
1968	1,752,000	1958	328,000	1948	628,840	1938	43,739	1928	23,969
1967	1,538,435	1957	224,000	1947	529,163	1937	40,920	1927	20,190
1966	1,200,000	1956	133,061	1946	367,946	1936	36,856	1926	1
1965	1,100,000	1955	97,000	1945	302,459	1935	34,336		
1964	1,008,889	1954	23,000	1944	269,561	1934	30,823		
1963	824,000	1953	855,726	1943	230,878	1933	29,562		
1962	744,000	1952	726,639	1942	143,509	1932	29,132		
1961	643,153	1951	709,249	1941	106,047	1931			
1960	516,000	1950		1940	99,775	1930	23,186		

A THOROUGHLY MODERN, VINTAGE WATCH DEALER

Scott Baratz, Boston MA, USA

Understanding your watch dealer is important, and goes a long way to ensuring you get a fair price on a good vintage watch. The adage that you should "buy the seller" is based on the notion that an experienced dealer has been in the business long enough to acquire a reputation worth protecting.

One such up and comer with a growing reputation is Scott Baratz, founder and CEO of VINTAGERLX Inc. Scott has been collecting and dealing in vintage watches with a special passion for the Rolex brand for over 20 years. He works around the clock to assist his clientele, connecting the right buyer with just the right watch.

He was initially inspired by his father's gift of his first Submariner, and the excitement and enthusiasm continues to this day. While Scott has an encyclopedic knowledge of Rolex reference numbers with a special love for Submariners and Sea-Dwellers, he is passionate about researching, learning and acquiring chronographs and dive watches from the 50s–70s.

Scott is emblematic of the next generation of vintage dealer who works to source watches from original owners and match them to the ardent collector.

"We're proud of the quality of our inventory and have built a reputation around it. We go to great lengths to source the best watches and parts, often from first owners. Buyers seldom know what goes into assembling top-flight inventory and it's a much bigger undertaking than simply selling them."

Scott travels the world sourcing vintage watches and matching them with buyers, which include both new and elite collectors, as well as other dealers.

"It's important to recognize dealers have overhead to cover and learning to be a good client is central to building trust. We place a high value on knowledgeable clients but also welcome new good natured buyers who have questions. We enjoy working with decisive, realistic people who know what they're looking for and understand the resources required to obtain what they want."

With this in mind, potential buyers should take the time to educate themselves as much as possible. Much of Scott's current inventory is displayed on the web at vintagerlx.com and through Instagram @vintagerlx.

Chapter 05 CONDITION

The condition of a watch is the primary driver of price, and both vary widely. When assessing the state of a vintage Rolex, you're looking for consistent, all-over, even aging and wear. If it appears uneven, with some parts looking new and others looking tattered, this is a red flag and price should be much softer. Uneven wear and inconsistent aging suggest unoriginal components.

Patina is a sign of aging and extends to the inside of the watch as well. Other indicators you should be looking for include corrosion and pitting, particularly around the opening, where the caseback screws down onto the midcase. Obviously, this applies only to stainless steel since gold will tarnish but not corrode.

If the case and internal mechanical parts look pristine with no apparent signs of age, you'd expect the dial and hands to be in similarly good condition and vice versa. Similar rules of thumb apply to bracelets, clasps, and even boxes and packaging.

If you inspect the movement, pay close attention to the state of screw heads. You're looking for consistency (all similar color and type). You do not want to see burring, which is minor damage resulting from removing and reinstalling screws with an incorrect or poorly maintained screwdriver. You might also see empty screw holes where a screw is missing. Hopefully, it's just been lost and isn't rattling around in the movement waiting to cause expensive damage. Even the most expert and careful watchmakers are prone to occasional mishaps.

None of the things described in this section should be deal breakers or prevent you from buying a watch you like. They are, however, useful indicators of value and price. If a watch has issues in ALL the areas mentioned in this section, it will – or should – be priced at the low end of the range, hard to resell and expensive to restore.

"The only thing I know is that I know nothing"

Socrates

WHY YOU NEED A GOOD SPRINGBAR TOOL AND HOW TO USE IT

If you want a vintage Rolex you need to examine what lies hidden beneath the lugs. You'd be surprised how agitated people get over this, and how unnecessarily complicated they can make it.

To get beneath the lugs, you need to remove the bracelet or strap. This is not difficult or technical or scary or dangerous to do. Nobody will get hurt and no permanent damage or harm will result. Just take your time. However, as a precaution if this is your first time – it's a good idea to apply some masking tape to the case lugs to act as a shield against any accidental slips and scratches.

Removing a bracelet or strap requires manipulation of the spring bars – the little springy pins that span the lugs and hold the bracelet (and end links) in place. Many find this easier if you separate the bracelet from the clasp so that it can lie flat.

You can do this with something as simple as a paper clip, or you can spend hundreds of dollars on specialized pliers made from surgical steel. Incidentally, the Bergeon tool (model 6825) is highly regarded and used by high-end jewelers. If you can afford the steep price and like to change straps and bracelets regularly, this sophisticated tool is a good buy and will make the procedure much faster and less frustrating. If you only perform this procedure occasionally, a simple spring bar tool will work fine and stretch your budget to only a few dollars. However, it's true that the spring bar tool takes a little practice and some dexterity to become proficient.

If your watch has lug holes, it's pretty obvious what the procedure is and how to go about using the tool. Holding the watch head firmly in your non-dominant hand, carefully insert the pointed end of the tool into the lug hole and press firmly.

If your tool has a fork-shaped end, do not use it on your watch! Use the pointed end only.

Insert the tool into the lug hole and press the tip of the spring bar pin all the way (about the depth of the lug hole). If the spring bar is old (with a rusty spring) you may need to press quite firmly to get it to move. In extreme cases, a small droplet of oil can be applied. This is a watchmaker's sized droplet, not diesel locomotive sized droplet!

When it does move, you may need to apply pressure to the end link (or strap) sufficient to prevent the pin from springing back. With pressure still applied, use the tool on the other lug hole to press in on the other end. You should then be able to work and wiggle the bracelet free.

If the case does not have lug holes, you will need to tackle the pins from the underside of the case. This is a more contemporary style that appeared in the 1990s.

You should see a gap in the end link exposing the spring bar. These are shoulder-style spring bars with a small notch in them. This is what the forked end of the spring bar tool is for. You will want to use the fork to grab the notch and pull it back. Apply some pressure on the bracelet or end link to prevent the pin from springing back into place, then repeat on the other side.

With both ends pressed, there is nothing holding the bracelet in place and you should be able to remove it. If the watch and pins are very old and dirty, some firm wiggling may be needed to work it free.

Bergeon springbar tool

PUTTING IT ALL BACK TOGETHER AGAIN

This is actually the more difficult part of the procedure. Can you see the two holes on the inside of the lugs? These are the positions you'll need to fit the spring bars back into. Before you begin, take a moment to make sure you're avoiding the big rookie mistake of installing the bracelet (strap) backwards – not upside down, but backwards. Check that the coronet on the clasp or tang buckle is oriented in the same direction as the coronet on the dial.

On the wrist, the clasp should close towards you. In the case of a strap, you will want to attach the length with the holes on the 6 o'clock side and the buckle length on the 12 o'clock.

Carefully thread the spring bar through the bracelet and end links or insert them into the strap. Insert one end of the spring bar into one of the holes on the lugs. In the case of aftermarket spring bars, it may not fit or reach. That's OK, as long as it's close.

Take the other end of the spring bar and compress it. You can use the tool or just a fingernail. With the pin pressed in, push the bar downwards into place. Apply pressure to the end link or strap and push it firmly into place.

It may require a wiggle to get the bars to spring into the holes. Listen for a firm click when this happens, then repeat on the other side.

If you don't own a spring bar tool, you really should. They come in various sizes and I find the smaller pocket-size to be the most practical and convenient. If you have large hands you might find a larger tool suits you better.

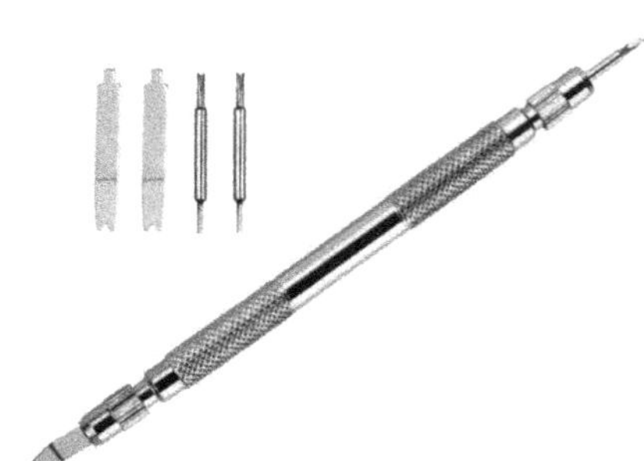

CASES

Rating the condition of a vintage watch case is subjective and there is no standardized scale. Over the years different case makers have supplied watch cases to Rolex. These have included C.R. Spillman (sometimes spelled Charles-Rene Spielman) and Genex, which Rolex acquired in 1998. Cases from both suppliers were sent to be finished and polished by a company called Joi Poli, which Rolex also bought the same year.

Vintage watches experience disassembly and cleaning several times during their lifetimes. Such servicing nearly always includes case polishing. A polished case suggests a watch has been well cared for by a conscientious owner.

However, collectors consider polished cases undesirable, and this is reflected in prevailing prices. For newcomers, this is counter-intuitive and confusing. There exists an unrealistic and idealistic pursuit of original mint condition vintage watches. While such examples exist, they are seldom seen in the wild and change hands with high price premiums.

It is essential to understand that polishing a watch case is a skilled craft. An unskilled watchmaker can ruin a watch in seconds and should be avoided. However, a shiny watch case shouldn't be mistaken for watches that have been repeatedly, but skillfully polished. Recognizing and distinguishing between the two takes practice.

Professionally repaired, restored and refinished cases that show as good as new

Elite collectors are willing to pay auction house premiums for top condition cases. These are investment grade watches that will never experience a sweaty wrist. Novice collectors hear the mantra that condition is everything and conclude that only mint condition and unpolished is desirable or collectible. This is misguided.

An average to a well-polished case should not put you off a watch you like. A poorly or over-polished case can be expensive to restore, near impossible to reverse, and will impact any future claims of originality and authenticity. The line between average and poor can be wide and is in the eye of the beholder.

The unhealthy obsession for like-new cases with their crisp chamfers (beveled edges) and satin finish has fueled the demand for refinishing services (known as case milling or lapping). These have become increasingly popular and available at a lower and lower cost. The craftsmen (and women) offering these services are skilled industrial engineers, proficient in computer-aided design and computerized lathes and milling machines. It is a highly professional undertaking, which can precisely mill (re-cut) a case back to its original form, complete with correct satin or even exotic finishes.

By combining milling techniques with modern advances in laser welding (reapplying metal), a heavily distressed or over-polished case can be made to look brand new and indistinguishable from New Old Stock (NOS) – old inventory, never sold, but changing hands amongst dealers as new.

The vintage Rolex market is now so heavily polluted with refinished cases that collectors should be highly skeptical of barn finds and claims of being all original, unpolished, or NOS. Without documented provenance, these claims should not command a price premium.

Social media, (Instagram in particular), is full of vintage watches with perfect cases paired with ghost bezels and deeply patinated dials. A ghost bezel is heavily weathered and faded. If you like the look, there are plenty to choose from, but be warned – the appreciation for these strange looking watches could pass as quickly as it emerged and these watches could be destined to become tomorrow's parts donors.

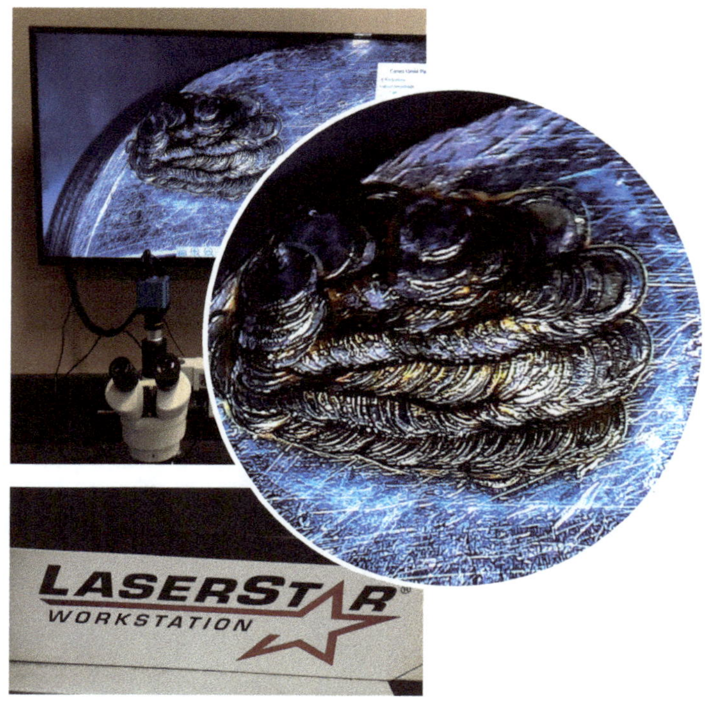

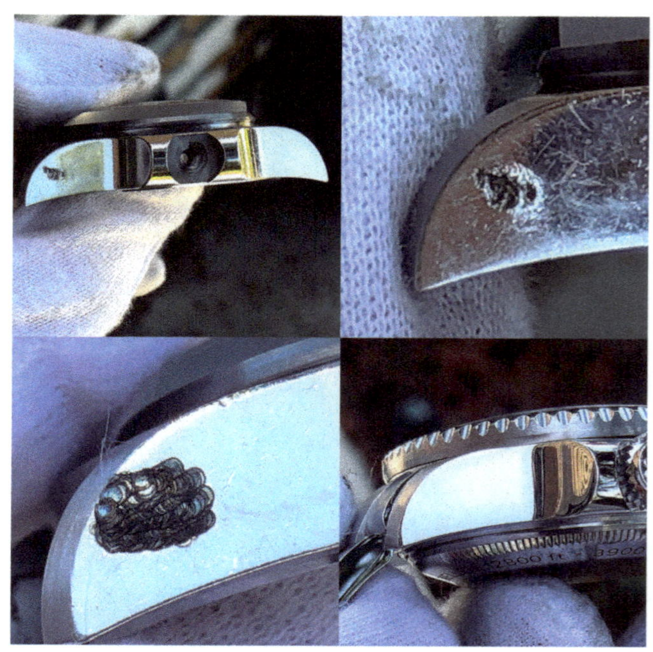

Benchtop laser welding has been common practice in the jewelery industry. Similar techniques are being widely used in vintage watch restoration and case repair.

CASEBACKS

Casebacks are interchangeable between Oyster cases of the same size irrespective of the reference. For example, a caseback from a 40mm GMT-Master will fit a 40mm Explorer, and it will not affect waterproofing, so it's not uncommon to see mismatched cases and casebacks.

Oyster casebacks usually do not feature any markings on the outside (the Sea-Dweller and some Tudors being exceptions). Rolex left them intentionally blank to allow owners to personalize them with engravings. These personalized engravings can be removed by skimming or polishing the caseback, but success depends on the depth of the engraving and the skill of the watchmaker. An engraved or personalized watch can be charming but difficult to resell unless it plays a role in the provenance of the watch (such as belonging to a celebrity).

Clear or display casebacks were only used on one reference, the modern (but discontinued) Cellini Prince. Any examples you encounter with clear casebacks will be aftermarket additions or pocket watch conversions (marriage watches).

The inside of a caseback will feature various stamped markings. These vary by reference and have evolved over the years. Typically you can expect to see the Rolex maker's mark, a model reference number, and in some cases, a date of production, and in others, a serial or issue number.

PROFILE OF A MODERN RESTORER

Mike Hui, CA, USA

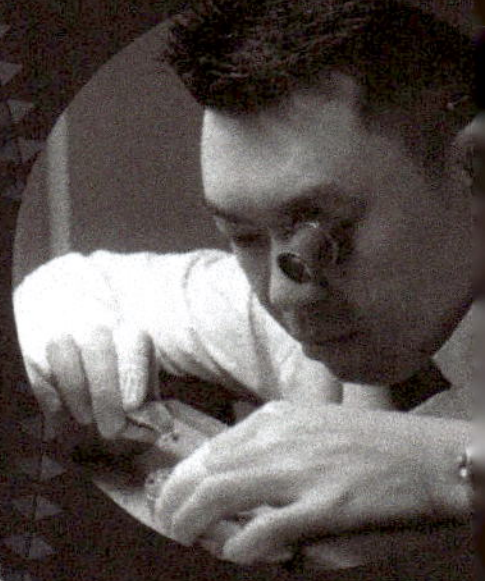

Every once in a while, an entrepreneur emerges to disrupt an established industry. One such professional is Mike Hui. Mike has been in the vintage Rolex industry since 2015 but in recent years has come to prominence for his highly acclaimed work in vintage case and bracelet restoration.

While movement servicing is an essential part of his business, he and his team have become highly acclaimed for the remarkable transformations they can achieve through advanced laser welding, milling, and polishing. While not unique in this field, they are recognized as the best in the USA, as much for their results as their customer service.

"Oyster cases can be uniquely challenging to work on due to their compound (parabolic) curves. Particularly on lugs. This requires lapping be performed by hand and assessed by eye. There is a very fine line between perfectly lapped and utterly destroyed and it takes experience to walk it."

Their work can be controversial among traditionalists, mostly because it is almost undetectable. Being able to return heavily distressed vintage watches to NOS condition makes any obsession with originality seem almost quaint. Mike goes to great lengths to emphasize there is no malicious intent to deceive collectors or manipulate values. Instead, the highly skilled and artisan nature of the work positions it alongside personalizations and customizations from the likes of Bamford and others.

Mike caters to collectors who understand the nuances between charming patina and outright tatty. Many engage his team to breathe new life into a beloved heirloom, setting it up for another century of regular wear and appreciation.

"For many of our clients there is a certain excitement about owning a piece of Rolex's storied history. It acts as a snapshot of where Rolex was at the time. For others it can also represent an important time period in their own life. I think it's very natural to imagine what the piece would have originally looked like. There's a comfort and pride of ownership in seeing their watch with the beautiful lines and edges the way Rolex intended."

Like other young and emerging influencers, Mike has a broad-reaching professional and social network. He has used this to benefit clients, facilitating transactions thanks to his remarkable access to some of the most desirable models. He also has the loyal patronage of many elite collectors and dealers.

If you're in the market to rescue a distressed vintage Rolex or need help procuring one, Mike can be reached online at rolliworks.com

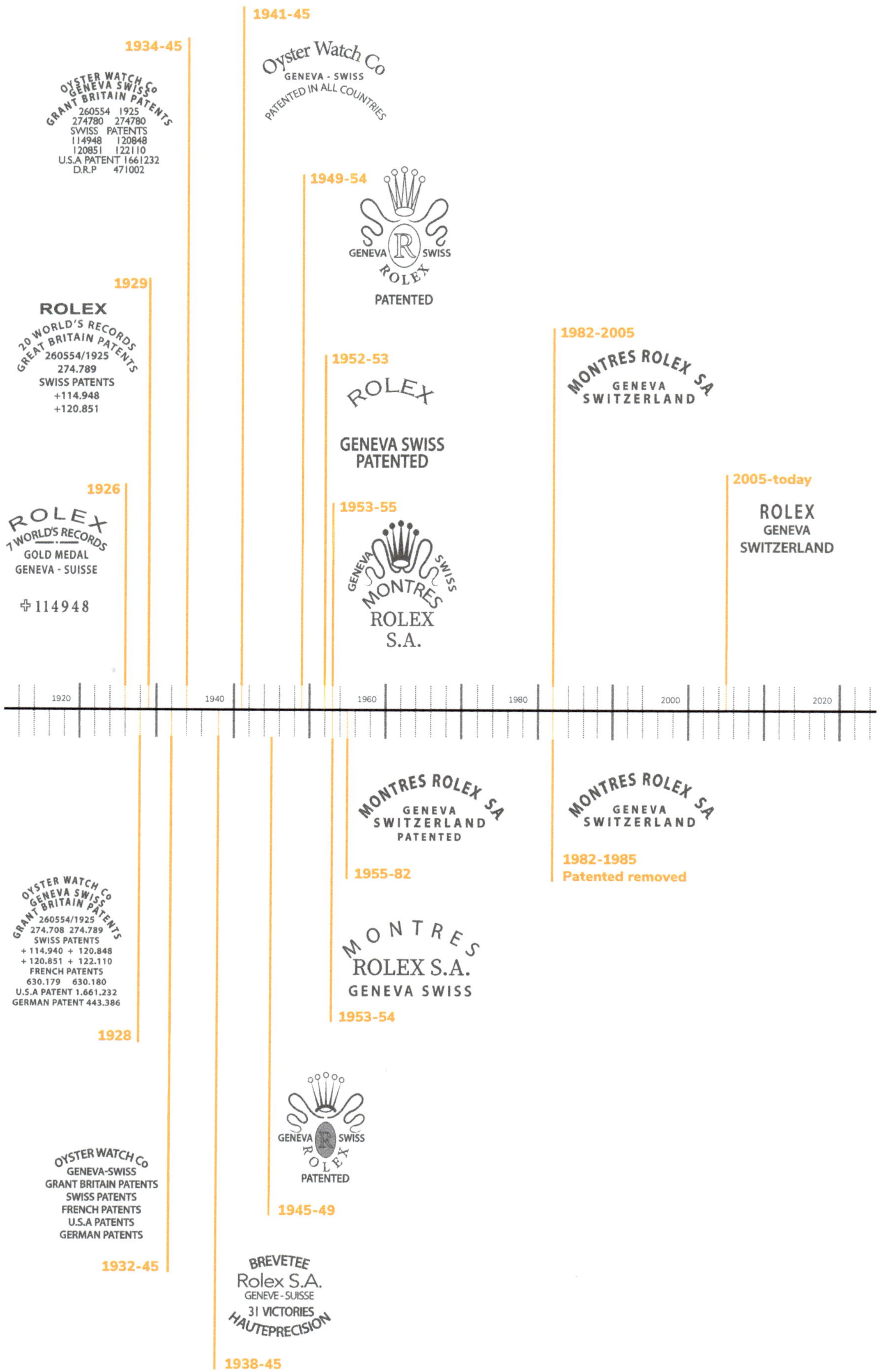

Caseback interior stamp markings

LUGS

The primary way to evaluate case condition is to look side-on at the position and shape of the lug holes. These spring bar holes should appear crisp and centered vertically. If they appear off-center (disproportionately close to the top edge), the case has been aggressively polished and only laser welding can add this lost material back. Lug holes should also appear round (not an ellipse or conical) and consistent on all sides. If they are splayed, off-center or uneven then you're looking at a bad polish.

The visual appearance of lugs wants to be consistent – meaning that the width and curvature should be uniform. While the top and bottom lugs won't be perfectly symmetrical, they should appear even in length and width. This can be difficult to judge as the Oyster case profile features numerous compound curves.

Not all case lugs had a chamfered edge. The presence and degree of chamfer varied by model and even within a reference. The edges and corners of lugs should be sharp and crisp, with a clear contrast between the satin and polished surfaces. Curved and rounded lug edges are the clearest indication of past polishing.

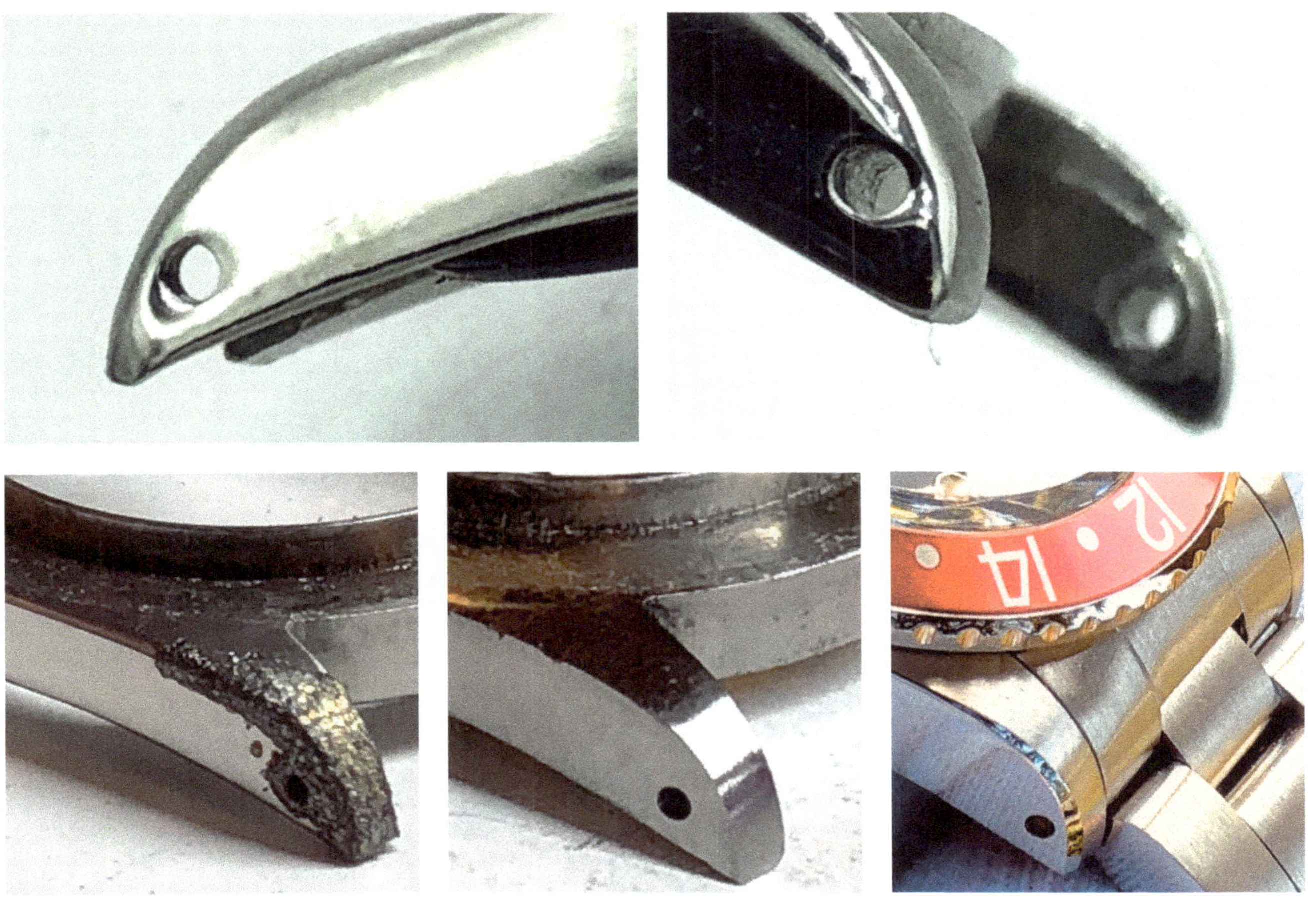

Top: Over-polished damaged lugs. Excellent candidates for restoration through laser welding and milling
Bottom: After laser welding, then recut (milled) with a chamfer.

FINISH

Another essential thing to look for is the case finish – especially satin and polished surfaces. While the finishes are subtle, Rolex went to great lengths to apply them.

You should be looking for signs of the different finishes, as they may well have been worn or polished away. If the finish is still visible, is it correct for the reference? Most vintage Rolex Oyster cases have a satin finish on the top of the case and a high polish on the sides. This is typical and most common. A few particular references such as the 6518 have this reversed with satin sides and polished surfaces. Only careful and diligent research can help you decide what's original and correct for these obscure exceptions.

On some references such as the Datejust, the sides of the case are bowed (convex) while others like some of the Oysterdate (e.g. ref. 6518) had square, straight sides. You should determine which is correct for the reference you're looking at and note whether these cues are still visible.

On some precious metal Datejust and Day-Date models, you can expect to find exotic finishes such as Bark or Morellis (e.g. ref. 1806). These hide wear and can be difficult to restore if damaged. While an acquired taste, they are rare and highly coveted and collectible.

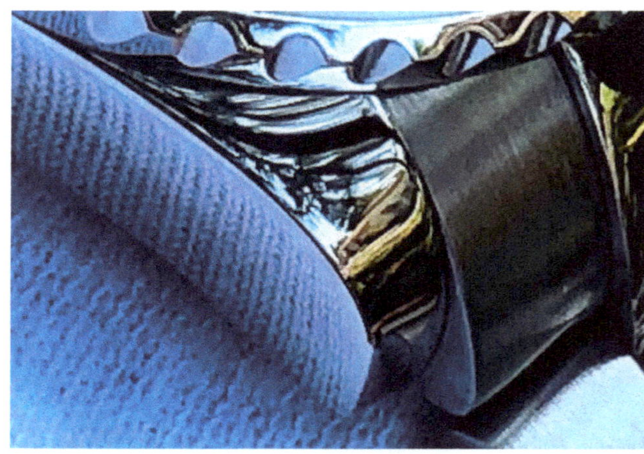

Top: Chamfered edges are individually polished to leave a crisp border with the satin, brushed finish to the top surface.

Middle: Chamfers have been eliminated on modern references like the Yacht-Master, which feature a high polish finish on all surfaces.

Bottom: The polished and satin finish on a bracelet should be correct, even if it is not standard issue with the watch (Jubilee was not an option for the Submariner)

CORROSION

The stainless steel used in Oyster cases is an alloy recipe made from iron ore, chromium, nickel, molybdenum, and a few other trace elements. These trace elements give the iron ore its stainless, corrosion-resisting properties. Metallurgists call this alloy "300 Series Austenitic Stainless Steel". 316 stainless steel is considered standard marine grade stainless steel. Despite the name, it is not resistant to warm seawater corrosion (specifically, chloride, fluoride, iodide, and bromide). 304L and 316L (316L is the low carbon version of 316) are derived from this 300 Series and are used in many pre-1987 vintage Oyster cases. Later Oyster cases use 904L steel with subtly different properties.

The common misconception is that 904L is harder and more scratch-resistant. The fact is, 316L has a higher Rockwell Hardness Rating (HR B 95) than 904L (HR B 7090), and 316L is also more scratch resistant and tougher to mill than 904L. It's said to hold a better polish as well. 904L does, however, have better salt and acid resistance.

In the mid-1980s Rolex adopted 904L as a solution to pitting and crevice corrosion. This is a common problem on Oyster mid-cases of sport and tool watches. It was first used on the transitional Submariner Date ref. 168000.

Urine attacks all austenitic stainless steel, so don't pee on your watches! Also, be sure to rinse them thoroughly after swimming to remove any chlorine. Bromide is just as harmful so keep vintage watches out of the hot-tub and away from the saltwater spa.

It's said that stainless steel needs to breathe and that regular cleaning with fresh water is necessary. When two pieces of stainless steel touch one another (such as a caseback and mid-case, or mid-case and bezel ring), cleaning and breathing is inhibited. Despite the trace elements that provide the stainless properties, corrosion will begin. Waterproof gaskets can accelerate this by trapping and retaining salts and oxidants, which, with time will trigger a galvanic reaction.

Like seawater, human sweat is highly acidic and rich in halides. Osmosis and capillary action will draw these corrosive elements into the case. With repeated exposure and evaporation the concentration of halides and oxidants accumulates and concentrates, giving rise to a reaction that compromises and penetrates the stainless protection offered by the trace elements.

This process is a chain reaction that is almost impossible to halt once started. It is observable as pitting and can only be prevented by grinding out the pit and filling the hole with laser welding. Even this drastic intervention is controversial, as it may not entirely halt the corrosion and may only serve to weaken the geometric lattice structure of the surrounding steel (through heat expansion and cooling contraction).

Pitting is bad news; it will shorten the life and destroy the value of a vintage watch. It is a slow and progressive chronic condition, and while some collectors enjoy the Wabi-Sabi aesthetic, others consider it cancer. While it shouldn't prevent you from enjoying the life left in a beautiful vintage watch, you should know that it will negatively affect longevity, collectability, and value.

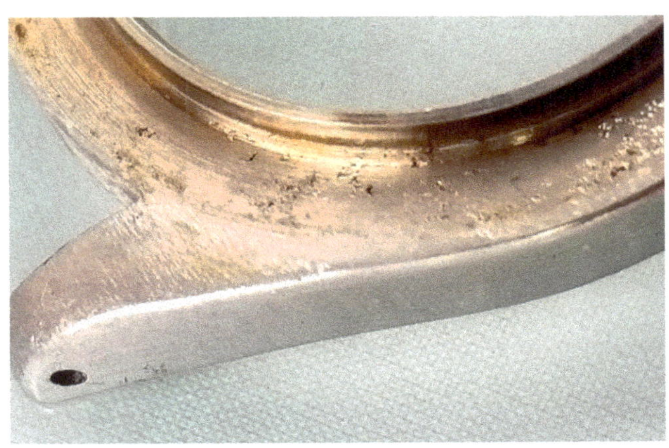

Early stages of pitting commonly appear on the top of the mid-case beneath the bezel ring. This is a challenging area to clean without disassembling the case.

HALLMARKS

Hallmarks can be complicated and intimidating, but for antique and vintage Rolex watches, we're really only concerned with six different hallmarks and fewer, if we exclude silver-cased antiques. After 1995 things get even more straightforward.

If you're in the market for a Rolex in precious metal, hallmarks matter. If a watch is presented and priced as 18K gold, hallmarks are your best way to validate the claim and ensure the entire watch, case, caseback, bracelet, and clasp are all 18K.

Hallmarks should be present on all Rolex made of precious metals. They appear on mid-cases (underside of lugs), casebacks, bracelets, and clasp blades. The absence of hallmarks indicates gold cap, gold fill or gold plate. Mismatched hallmarks suggest a Frankenwatch assembled from various parts.

Hallmarks are defined by an official Swiss government assay office and applied under license by the manufacturer (Rolex). There are seven assay offices in Switzerland and Rolex uses the Geneva office exclusively. The symbol for the Geneva assay office is a capital G, which will form part of a larger hallmark.

Antique and vintage Rolex watches (pre-1995) have a hallmark symbol for each different purity of gold and platinum (9K, 14K, 18K). However, there is no distinction between different colors of gold (white, rose, yellow).

A Swiss government hallmark is exceptionally detailed, which makes it very difficult to copy. It must be consistent and flawless with each stamping. The assay office controls the quality and if not perfect, the item will be scrapped and cannot be sold. The crispness and detail of a hallmark are essential in establishing authenticity.

Solid 18K (750) gold watches will have the Helvetia Bust (lady's head) with a capital "G" below her neck. 14K (583) gold will have a squirrel with a capital "G" over it. Platinum will have an ibex with the "G" between the antlers.

Barry the St. Bernard Hallmark in this illustration has an X marking the position of the capital G for the Geneva assay office

Hallmark symbol styles evolved and in 1933 a morning star (Morgenstern) was introduced for 9K (375) gold watches.

Current Swiss law (post-1995) has reduced the number of hallmarks to a single mark, the head of a St. Bernard dog. This official hallmark is used for all precious metals in all qualities. The Office of Precious Metal Control refers to this St. Bernard dog as "Barry". It appears near the Maker's Responsibility Mark and their indication of purity.

Barry was of a breed (Bernhard) which was later called the St. Bernard. He worked as a mountain rescue dog in Switzerland and Italy for the Great St Bernard Hospice. He has been described as the most famous St. Bernard and credited with saving more than 40 lives during his lifetime. His German name is Menschenretter, meaning "people rescuer."

Rolex watches imported into other countries are subject to local assay office inspection and may carry hallmarks in addition to the Swiss ones. For example, watches imported for sale in France and England would have additional, but different hallmarks. National hallmarks are numerous and out of the scope of this guide.

ANTIQUE HALLMARKS PRE-1933

Gold		Silver		Platinum (from 10.02.1914)
750‰	583‰	875‰	800‰	950‰
Helvetia	Écureuil	Ours	Coq de bruyère	Tête de chamois

ANTIQUE HALLMARKS POST 1933

Gold			Silver		Platinum
750‰	585‰	375‰*	925‰	800‰	950‰
Helvetia	Écureuil	Morgenstern	Canard	Coq de bruyère	Bouquetin

* only for watch-cases

ASSESSING DIALS

This is the most critical, confusing, and technical aspect of evaluating a vintage watch. A dial can account for 50 to 100 percent of the value premium between watches of the same reference. A dial's condition, originality, distinctive marks, original components, and the degree of re-touching can be the difference between priceless and worthless.

Rolex has used several suppliers for dials. These include Beyeler, Singer, Stern, Lemrich, and Metelem. Each stamped their name on the reverse side of the dial. Some of these marks are only visible under powerful magnification.

Small distinctions such as a retailer's co-signature, a military mark, or manufacturing defects can produce large price differentials. It is essential to look out for underhanded and dishonest practices concerning these descriptions. Buyers often confer with other collectors or consult forums to assess the appropriateness and accuracy of a seller's description.

If you are pursuing high-value references, it is essential to consult authoritative published sources. A specialist, technical volume exists for almost every vintage reference. Most will credit vetted sources of photographs and information, including acknowledgments to significant collectors, factory archivists, and museum staff. Recognized authoritative publishers include among others, Guido Mondani, Pucci Papaleo Editore, James Dowling, John Goldberger, and Osvaldo Patrizzi.

Dial manufacturer Singer would stamp their makers mark on the back of the dial plate. The style of this mark evolved over the years.

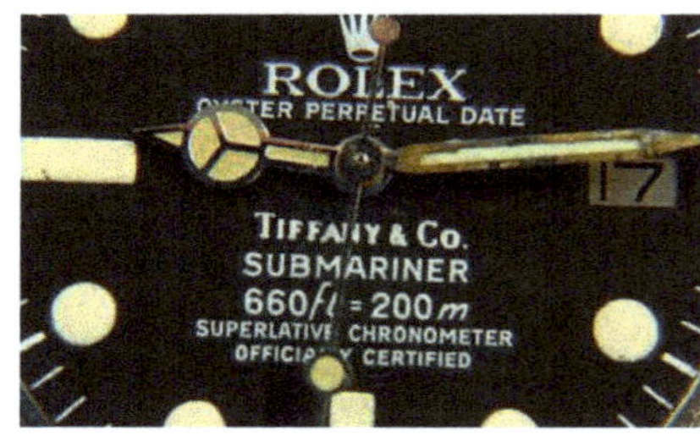

Tiffany and other famous jewelers have a long history of co-signing dials.

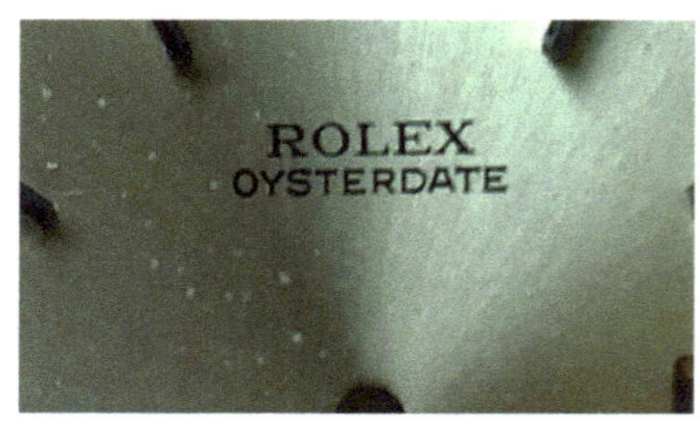

Water damage is commonly seen.

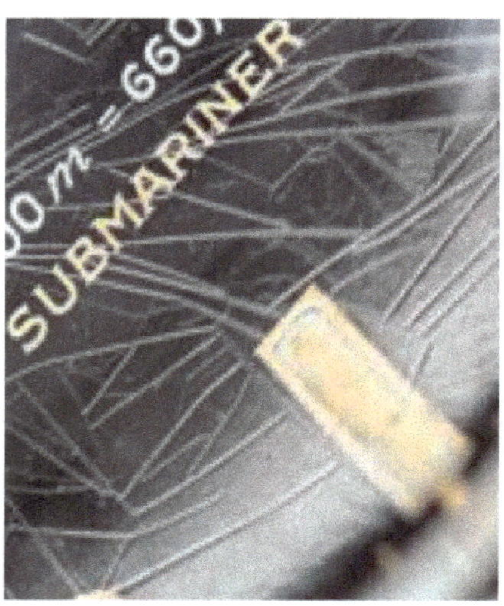

Lacquer defects can result in eventual cracking, giving rise to the creatively-named spider dials. If uniform and symetrical they can be attractive but the long-term prognosis is not good. If left untreated and stablized the lacquer will eventually chip and flake. Caught by moving hands it will scratch the dial leaving drag marks, before being ground down and making its way into the movement.

The Beyeler 'B' stamp

DIAL PRINT

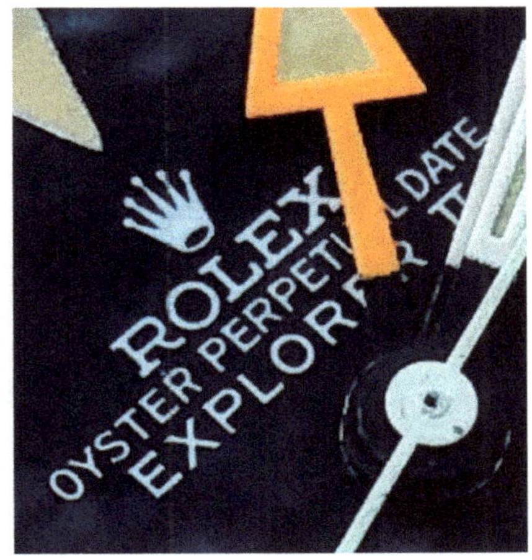

Manufacturing technology during the mid-20th century was not perfect, and printed dial text can vary significantly from the start to the end of the model run. Collectors can expect minor and subtle differences in font, color, and texture of printed text.

Except for some late Modern Classics, dial production was outsourced to partners like Singer and Beyeler. Different batches of dials had minor variations in fonts, print quality, color consistency, finish, and texture.

Gilt dial text (bronze color) was printed by a process of reverse osmosis, revealing the brass base plate below the paint. While finished with a clear coat of lacquer, the brass can still darken with age. These are very desirable and were made until the late 1970s. These gilt dials are also the subject of sophisticated counterfeiters with very high-quality fakes in wide circulation. There are poor fakes, but these days, there are increasingly "Super Fakes" as well. These are almost impossible to spot without disassembly and forensic examination.

The top row of dial text (e.g. Rolex, Oyster Perpetual) often has a pinched or etched appearance that differs from the bottom row of text (e.g. Officially Certified Chronometer or the depth rating). The bottom row, or specification text, was commonly printed later in a separate production process. This allowed generic dials to be branded first, then customized for different models and movement grades. It also means there may be inconsistent fonts on the same dial.

The Rolex coronet on dials has also evolved and shows many variations. Some are reference-specific while others are period-oriented. The details described in this guide attempt to capture the catalog configurations, as illustrated in marketing materials at the time of launch. It is known that dial print evolved throughout the product's life cycle, deviating from the original first batches of watches.

All the possible variations from catalog configurations cannot be listed in this guide due to their sheer number and a lack of an authoritative way to verify their correctness.

Bubblebacks and Explorers 6610 and 1016

Submariners 6205, 6200, 6538, 5510 & GMT-Master 6542

Submariners 5512, 5513 Gilt, 5513 White, & 5513 Bart Simpson

Submariners 1680: MK1 Red, MK1 White, MK2 Red, MK4 Red, MK5 & MK6

Sea-Dweller 1665: DRSD MK2, White MK1, MK2 Rail Dial, White MK4, Single Red Service

DIAL LUMINOUS

There are five distinct types of dial lume used throughout Rolex history, each with unique characteristics (radium, tritium, Luminova, Super Luminova, and Chromalight). It is common for collectors to use a Geiger counter to assess the originality and authenticity of antique and vintage lume. This is in response to the increasingly high quality of counterfeit dials using degraded radium and tritium lume material.

The first generation of lume was radium, used in the 1920s to 1940s. This was highly radioactive and still poses a health risk today if accidentally ingested (i.e. by inhaling radium dust). As the dangers became apparent in the late 1930s, watchmakers attempted to reduce both the amount of radium applied to dials and the radioactivity by diluting the formulation.

When exposed to moisture and humidity, radium lume will age to a dirty green color, resembling mold. With time, radium will burn dials and corrode crystals. This cracking effect on acrylic crystals is called crazing.

The second generation of less dangerous radium was introduced in the 1940s and used in particular, on the earliest Submariners, Explorers, and Turn-O-Graphs from 1953 to 1956.

Tritium emerged in the early 1960s. The luminous material isn't pure tritium but a phosphorus compound, which uses tritium as the energy source that causes it to glow green. The tritium has a half-life of 12 years (meaning zero glow after 12 years).

Depleted tritium (actually, the phosphorus in the tritium compound) will respond to UV light by holding a glow for a few seconds. Several tritium recipes and application techniques were tried over the years affecting the duration of the glow.

Before 1965, tritium was applied by hand, resulting in a raised puffy appearance. After 1965 it was pad printed resulting in a flatter, smoother appearance.

Lume type by six o'clock dial markings

Aged tritium on a vintage Turn-O-Graph circa 1950s. Note the cream hour markers and the inconsistent 6 o'clock marker. Overall originality is inconclusive but most likely a vintage-relume.

A typical moisture-damaged tritium dial and hands from a GMT-Master 1675 circa 1960s. Note the almost moldy appearance of the lume.

Radium burn on a vintage Bubbleback 2940 circa 1940s. The lume has assumed a distinctive olive green hue.

A GMT-Master 1675 circa 1960s. Note the perfectly aged dial lume is inconsistent with the hands. This is not uncommon and not necessarily an indication of a re-lume.

From 1983 to 1997, tritium lume was edged with white gold surrounds. These are concave hour markers designed to hold the lume material in place. They prevent spreading, smudging, leaching, and chipping.

Early hand-applied tritium responds under UV light today. Matte dials of the mid-1960s to 1970s had either the hand-applied raised puffy appearance or the smooth pad-printed appearance. The pad-printed style no longer responds to UV light stimulus.

By the late 1970s, lume had a look somewhere in between puffy and smooth – a uniform domed appearance. This lume doesn't respond to UV light but glows faintly in a dark room. By the 1980s the appearance had changed again, with a smooth glossy look. While it doesn't respond to UV light, it will still glow in dark conditions.

Exposure to moisture and humidity will cause tritium to turn unevenly black, and is considered unattractive and undesirable. Without moisture, it will age (patina) evenly to various shades of brown.

Luminova and its derivatives (Super Luminova and Chromalight) are considered thoroughly modern, and will still glow green and blue, respectively in the present day. Watches with Luminova and its derivatives are not radioactive and pose no danger at all.

Type	Dial Marking	Lume Material	Period
1	SWISS	Early Radium	1920s–1930s
2	SWISS	Later Radium	1940s–1963
3	T Swiss T	Early Tritium	1963–1999
4	T Swiss T<25	Early Tritium	1964–1967
5	T Swiss - T<25	Later Tritium	1967–1997
6	σ T-Swiss-T σ	Later Tritium	1970–1979
7	- T SWISS T -	Later Tritium	1970s–1998
8	SWISS	Luminova	1998–2000
9	SWISS MADE	Superluminova	2000–2008
10	SWISS MADE	Chromalight	2009–Present

PROFILE OF A MASTER DIAL LUMER

James Hyman, Northumberland, UK

Within the realm of vintage watch restoration there are few sub-specialties as controversial as dial lume. Uninitiated collectors shun all relumed dials, as they do refinished ones. Seasoned, experienced collectors have a more enlightened and nuanced perspective. The fear that a relume will destroy the value of an original dial ignores the fact a dial in need of a relume has already lost much of its residual value, and the net effect can only benefit the whole watch.

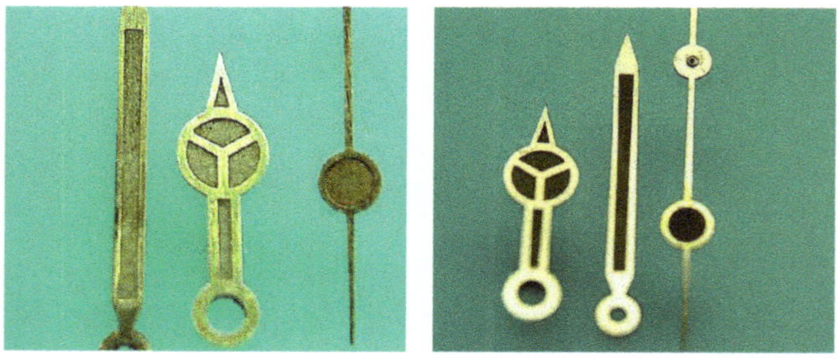

Skillfully relumed and stabilized hands

James Hyman is one of a handful of up and comers making a name for themselves in this highly skilled subspecialty. James has a low-key digital and social footprint, in part because of the controversy surrounding his craft. He has been reluming dials since 2010 and maintains a small but exclusive client list of ardent collectors. He is sought out by enlightened collectors seeking highly customized work on prized, high-value watches.

"Dial relume services exist for two reasons. 1) to stabilize deteriorating lume and 2) to improve cosmetic appearance. Both are important considerations when deciding to relume a vintage watch."

Aged luminous materials on dial markers and hands will corrode and degrade with passing years. Eventually it will crack and flake, detaching itself from anchor points. Loose flakes can scratch and dislodge peeling dial lacquer, leaving detritus between the crystal and dial. If left unchecked it will break down and enter the movement. This gritty compound will cause havoc with lubricants and gears.

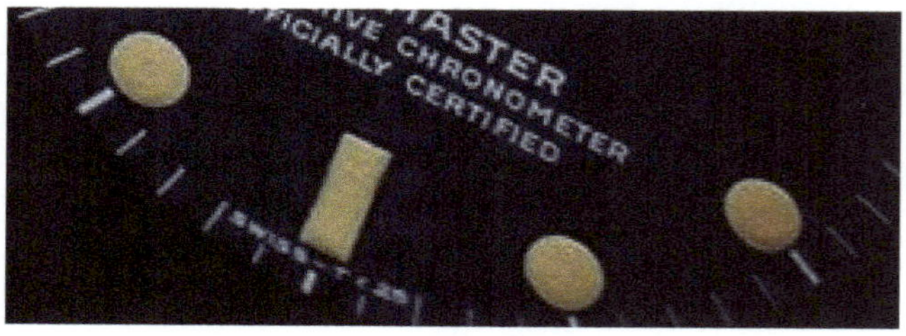

"The type of relume can range from Luminova to produce a factory-new effect to spent tritium for a faux-vintage look. A like-new relume will glow brightly and look brand new. A client seeking a vintage relume can specify the degree of brightness and the color and texture of the lume paste. Various formulations of aged tritium can be mixed to get exactly the right colour, texture and glow to match hands or just satisfy a discerning eye."

If you have a distressed dial and hands in need of stabilization, restoration, or just beautification, James can be reached via Instagram @thealchemistrelumer

THREE STEPS OF DIAL RELUME

STEP 1 requires removing existing lume using shaped peg wood (a sharpened dowel) or a small screwdriver. The technique is to scratch away only the lume paste without touching the dial surface beneath. Removing a previous relume can pose unexpected challenges if the lume recipe includes too much binder. This will result in a glossy appearance and may require pre-treatment to allow it to be broken down and removed.

STEP 2 involves cleaning the dial of any loose material to reveal the condition of the surface beneath. Any damage will need to be fixed to ensure a stable surface for new lume to adhere to.

STEP 3 involves applying the new lume with various sized oilers (small tools designed to precisely apply lubricant to movement components). In most cases, lume is applied in one sitting but complex numbered dials may take more. The trick is to complete application within the curing time of the lume mixture. Use of solvents to extend the curing time can affect the final surface texture and appearance. The mix formula, application technique and curing conditions will result in a unique appearance. So for a consistent and even final result, the task should ideally be finished in one sitting with the same batch of lume mix.

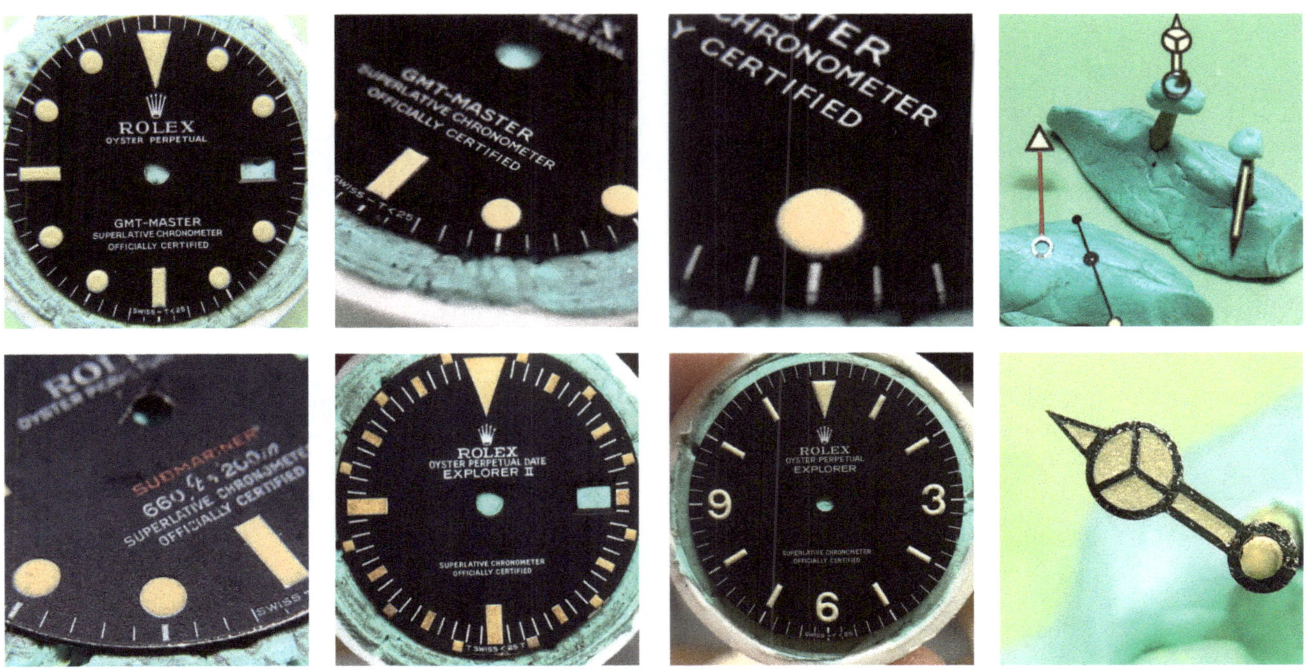

SIGMA DIALS

The characters of the type six dial (on page 46) are the lowercase Greek letter sigma (σ), giving the dials the name, sigma dials. The sigma symbols were intended to indicate that the hands and indexes on a watch were made of solid gold. The dash marks shown as type seven indicate steel indexes and hands.

The sigma character was in use throughout the 1970s. Gold prices tripled during the early 1970s, and the sigma dial was an attempt to emphasize the intrinsic value of the watch. The quartz crisis (arrival of very accurate, inexpensive quartz watches from the Far East) was underway, and several Swiss watchmakers adopted the convention in an attempt to highlight their luxe credentials.

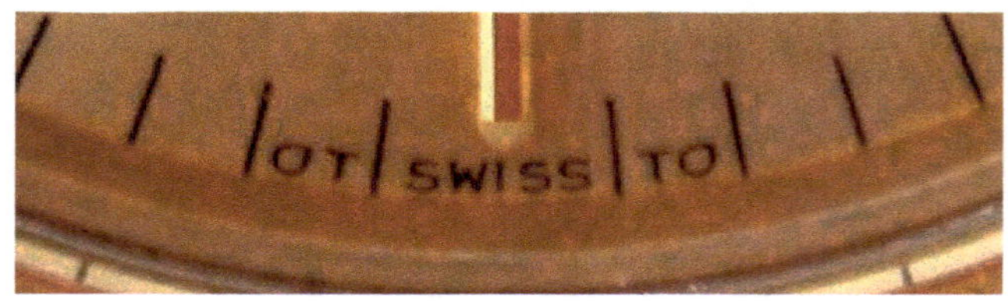

ARABIC SCRIPT DIALS

A class of dial exists that features Eastern Arabic script numerals. These are not to be confused with the Arabic numerals we have become used to seeing (1, 2, 3, and so on). These Arabic script dials feature symbols descended from the ancient Hindu-Arabic numeral system and are commonly referred to as, Arabic script. Today these are used throughout the Arab world and in particular the Gulf States.

Rolex first offered dials with Arabic script in 1956 in the Day-Date refs. 6510, and 6511. In 1957, Rolex added refs. 6611, 6612 and 6613 featuring fluted and diamond bezels.

Rolex pioneered watches with Arabic script dials with matching Arabic script calendar functions. Rolex was an early entrant into the prospering Gulf market during the 1960s and these localized dials proved popular.

The introduction of Arabic script was gradual, starting with the day of the week being spelled out in Arabic. Later references have Arabic script on the roulette date wheel too. Most of these Arabic dials appear in white gold or platinum, as Islam prohibits men from wearing yellow gold for reasons of modesty.

In 1961 Rolex introduced refs. 1802, 1803, 1804. The ref. 1804 acquired the nickname Scheherazade, after a female storyteller in the classic Persian collection of tales, One Thousand And One Nights. In opulent platinum, it had a diamond bezel and proved very popular, remaining in production for 20 years. Ref. 1803 in pink gold was available with a unique tapisserie guilloché dial and acquired the nickname, Aladdin's Rose. Phillips sold a well-documented example in 2015, which was made in 1974 and first sold in Damascus, Syria.

Throughout their production, these dials maintained consistent typography (font style and size) on the dials, and day and date wheels. They were eventually retired in the early 1980s but the Arabic script day and date wheels remained available as an option which could be swapped-in by an authorized dealer (or AD).

In 2016, Rolex reintroduced a platinum Day-Date 40 with a dial featuring full Arabic script. The ref. M228206-0025 has a pale blue dial with contrasting dark blue hands and Arabic script. Curiously, the typography appears to be the same as the original versions. Distribution is limited to ADs in Kuwait, Qatar, Saudi Arabia, and the territories of the United Arab Emirates.

١	٢	٣	٤	٥	٦	٧	٨	٩	١٠
10	9	8	7	6	5	4	3	2	1

ARAB CREST DIALS

A subgenre of Arab dials features crests (logos) of various Gulf States. The very first examples are thought to be special orders from the Sultan of Oman, Qaboos bin Said. These were first ordered through Aspreys of Bond Street in London. Later versions were ordered by Khimji Ramdas, an Authorized Dealer in Oman. These Rolex OEM examples have unique engravings on the inside of the caseback, which feature the case serial number and the name of the jeweler.

Crests for other Gulf nations were applied by local dealers, rather than by Rolex at their factory. Although less desirable, they are collectible as are other special logo-dials. Logo-dials feature all manner of corporate brands and were often awarded as long service or special recognition gifts.

Submariner bearing the crest of the United Arab Emirates for armed forces

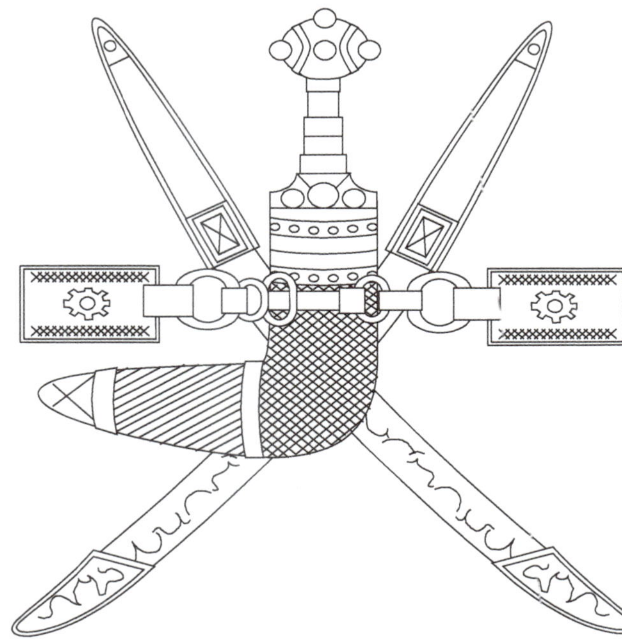

National emblem of the Sultanate of Oman, typically printed in red on the lower half of a dial

GILT DIALS

Gilt dials are some of the most desirable and collectible on vintage Explorers, Submariners and GMT-Masters. They appeared from the 1950s through the early 1980s. They are known to include variations on watches of the same model and close proximity in serial numbers (age). Many of these variations are sometimes mistakenly described as printing errors (error dials) and sometimes mistakenly deemed counterfeit. They are also the subject of many common misconceptions. For example, the gilt can't flake off, or is not gold or etched.

These variations are genuine and natural, and can result in the appearance of noticeably fatter or thinner fonts with sharper or blurrier serifs. These variations are a result of changing production tolerances of tools and materials that behave differently and produce varied results as the environment changes.

For example, printing pressure and duration will evolve with mechanical wear. Materials like lacquer and paint respond differently to seasonal changes in temperature and humidity. Printing pads are abrasive and elastic and slowly wear out with each application.

While these factors may affect the general appearance of gilt text, coronets, and chapter rings, they will not affect spacing, size, and shape. This means that an oval 'O' will not become more circular, and text and character alignment will not change, nor will spacing between lines.

Rolex has never revealed their exact production process. However, the production of gilt dials in general is known to involve printing onto the bare brass dial plate with a pad. The coronet, text, and chapter ring is printed with a clear lacquer. The dial is then immersed in a galvanic solution and the black paint adheres only to the exposed bare brass and not the printed clear lacquer. Once dry, the whole dial is given a clear top coat of lacquer.

Under magnification the text appears in relief to the paint

This final clear coat seals the dial and prevents oxidation and tarnishing of the shiny brass. It is this final clear coat that can be prone to flaking and peeling.

Even today with modern galvanic production processes it is extremely difficult to make a 100% identical replica of a genuine Rolex gilt dial. However, improvements in 3D printing technology enable high quality copies of printing pads derived from the lacquer negative. With convincing replicas in circulation, recognizing authentic variations becomes harder. The prevailing advice is to focus on the appearance of the texture of the gilt and lacquer surfaces and their general patina.

DIAL REFINISHING

Until about the 1980s, it was common practice during an independent service to refresh a dial with new paint or lacquer (Rolex Service Centers would replace them entirely). This often extended to reluming hands and dial furniture (hour and minute markers, Rolex coronet, and date window border). This was done by hand with results ranging from ghastly to convincing.

Like a refinished case, a refinished or relumed dial is viewed as undesirable. These are no longer considered original, despite often being quite attractive. Advances in technology have made results so convincing as to be indistinguishable from the original, complete with faux aging.

The process of refinishing begins with stripping a dial down to the bare metal plate. The dial is soaked in a solvent to remove all traces of paint and lacquer. Dial furniture, such as applied indexes and the coronet, should also be removed during this process, though this step is sometimes skipped for fear of damaging parts during removal and reattachment.

The clean base plate is then repainted and reprinted. Ideally, this will be done before reapplying dial furniture. If left in place, you may see paint on the sides of the applied hour markers. This is a telling indication of a refinish.

Artfully refinished dials can be beautiful. This 65-year-old example was faithfully returned to what it would have looked like when sold in 1953. There was no attempt to give it a faux-vintage appearance.

Paint appears to run onto the hour indexes. They were not removed due to a risk of damage to the dial plate and the markers.

A tasefully refinished dial

Refinishing is sometimes confused with redialing. A redial is replacing a dial entirely with say, a service dial, or a naturally more attractive one from a donor watch.

A vintage refinish is considered more acceptable than a modern relume, but both are viewed negatively by collectors and will detract from price and value. The same rule of thumb applies to authentic Rolex service dials – less desirable but still acceptable. A modern refinish that attempts to recreate faux-patina is highly frowned upon.

Collectors are typically on the lookout for dials that have been refinished or otherwise restored. They should also be on the lookout for dials that have been artificially distressed. A dial can be literally baked in an oven to increase or enhance the appearance of patina. Many Daytona 16520 Patrizzi dials have been subjected to this controversial practice. For this reason, only serious expert collectors should dabble in this reference.

DIAL MATERIALS

Many collectors are surprised by the sheer variety of materials used to make dials. Modern collections are more conservative with the use of exotic materials. Mother of pearl and meteorite dials are still in use on the higher end, precious metal product lines like the Pearlmaster, Daytona, and Day-Date.

The exotic dial materials listed below are known to be correct for vintage Rolex watches and appeared predominantly on Datejust and Day-Date models. Collectors should be wary of brightly colored dials, as most are aftermarket refinishes or outright counterfeits.

Agate	Ammonite	Aventurine	Azurite
Bloodstone	Cacholong	Carnelian	Chrysoprase
Blue Coral	Pink Coral	Yellow Coral	Ferrite
Lapis Lazuli	Sodalite	Tigers Eye	Sugilite
Fossil	Grossular	Howlite (Marble)	Green Jade
Lavender Jade	Green Jadeite	Blue Jadeite	Green Jasper
Rose Jasper	Malachite	Meteorite	Nacre (MOP)
Obsidian	Onyx	Opal	Rubellite

Exotic dial materials

HANDS

Over the years, Rolex has used several watch hand styles from various suppliers (including Universal, Fiedler, and Virex). They each produced a variety of different styles. Knowing the appropriate type for a period and collection goes a long way towards determining authenticity and originality. Newcomers commonly overlook the condition and correctness of hands when reviewing a vintage Rolex. They may also dismiss an authentic watch with unusual or rare hands.

A few important rules of thumb to consider:

- The style of hands should be consistent and complementary to the dial indices and coronet. For example, large and rectangular baton style hour markers typical of the 1980s should have similarly-styled large baton hands. Diamond-shaped or arrow-shaped hour markers characteristic of the 1950s would be paired with dauphine or alpha-style hands.
- The color and material of the hands should match the dial furniture, case, and bezel. You would not expect to see gold colored hands on a white dial or paired with silver hour markers.
- The state of corrosion should reflect that of the dial. Solid gold hands may tarnish but will not corrode, unlike the base metal beneath the gold plate. A solid gold watch would not have electroplated hands.
- Not all hands were lumed – even hands of similar style. For example, Alpha hands may or may not contain a lume strip, but they should still match the dial. If they have no lume, there should be no lume dots, strips or fills on the dial.
- Radium and tritium lume on hands will corrode and patina. It is uncommon for this to be an exact match to the patina of lume on the hour markers. Lume for the dial and hands would have been applied at different stages of production and would have come from different batches of material. It is the exception rather than the rule that they age perfectly evenly.
- Dress watches such as vintage Cellini generally were not lumed, but there are a few exceptions.

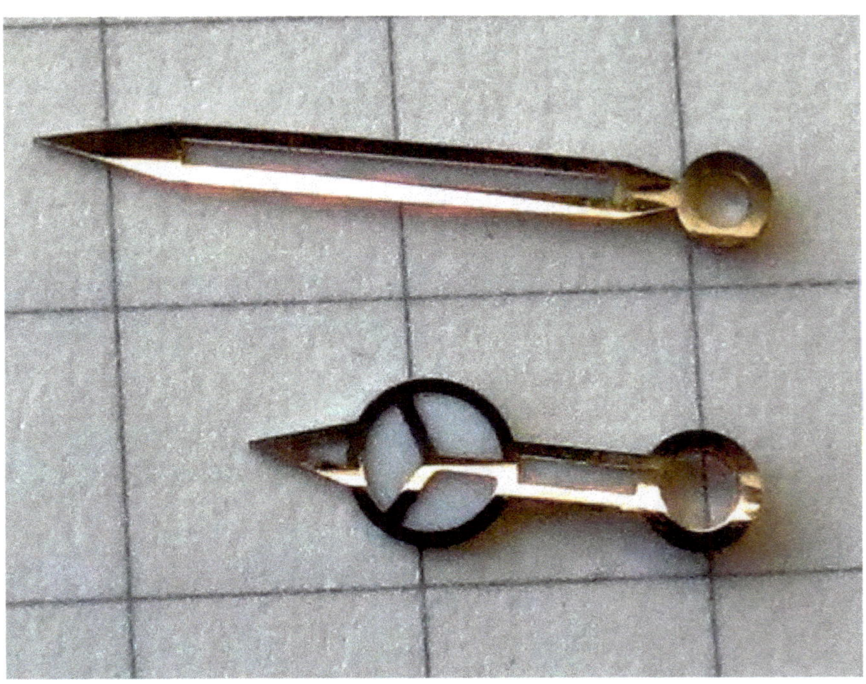

Contemporary and Service hands have a curved (convex) profile which reflects light differently from flat vintage hands. This example is filled with bright white Luminova which is the first indicator of age. You may encounter curved, modern hands that have been relumed with aged tritium.

Hands are made from a material that reflects the dial hour markers, which themselves complement the rest of the case and bezel. Hands may be electroplated, solid gold, or steel. They can be finished in brightly-colored paint or heat treated to become blued steel (seen on early seconds hands).

1. The formal name for Cathedral or Mercedes style hands is Squelette. They appeared on Rolex professional tool watches and were designed to hold more lume material. They are used only as hour hands.

2. Alpha hands have a broad base and narrow stem to attach to the mounting post at the center of the dial. The style is used for both the hour and minute hands and they often have a tritium lume strip. This style is typical of the 1940s to mid-1960s and seen on Bubblebacks, Datejust, Oysterdate, the later Oyster Date, and Oyster Perpetual.

3. Dauphine hands are often confused with Alpha-style hands. They are used for both hours and minutes and are very traditional-looking, triangular shaped and faceted. They usually appear on Bubblebacks, Oyster Dates, and Datejusts.

4. Syringe hands occasionally appear on antique Rolex pocket watches and early Bubblebacks. They are uncommon and usually have lost any radium lume material they may have held.

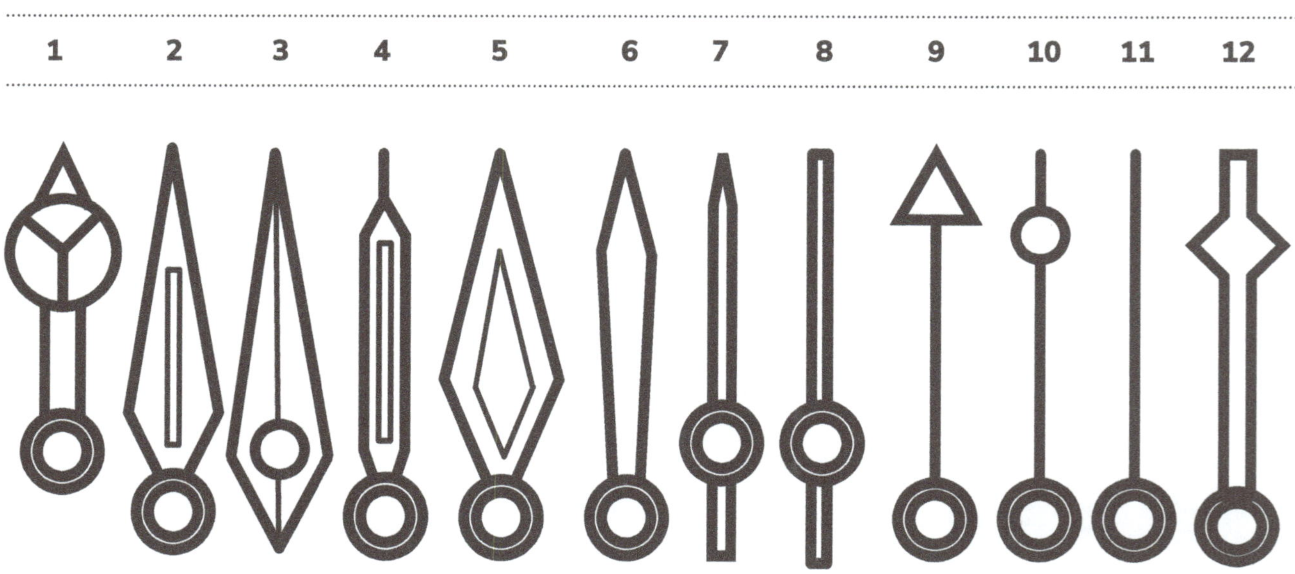

Stylized illustration of hand types used by Rolex

5. Plongeur Hands are high-visibility hour hands for diving watches. They are broad and diamond-shaped, appearing on military issue Submariners only. The accompanying minute hand may be a Pencil or Batten style.

6. Sword hands resemble a sword blade. They taper from the center mounting post and get progressively broader before narrowing to a sword tip. They may feature a lume panel of radium and are seen on Bubblebacks from the 1930s and 1940s.

7. Pencil hands are a cross between the Batten style and the Sword style. They have a straight body and narrow distinctly at the tip resembling a pencil. They were used widely in the 1960s and 1970s.

8. Broad Batten hands are formally called Parformes. They are straight, elongated and somewhat minimalist. They may or may not feature a lume strip and were used commonly on dress watches.

9. Arrow hands have a triangular arrow pointer on the end of a straight hand and are used for 24-hour GMT functions only. Variations can be seen in the size and color of the triangular tip. The Explorer II 1655 has a unique and unusual variation on this arrow hand and is illustrated later in this guide.

10. Lollipop hands are used for second hands. They are used on professional tool watches only. The lollipop is filled with tritium or Luminova for later watches. The Milgauss has a unique lightning bolt second hand not depicted here.

11. Narrow Batten hands are second hands, which are typically seen on classic and dress watches from Bubblebacks to modern Datejusts. These are paired with Broad Batten hands.

12. Snowflake hands are only used on Tudor Submariners and feature a unique square on the end of the hour hand only.

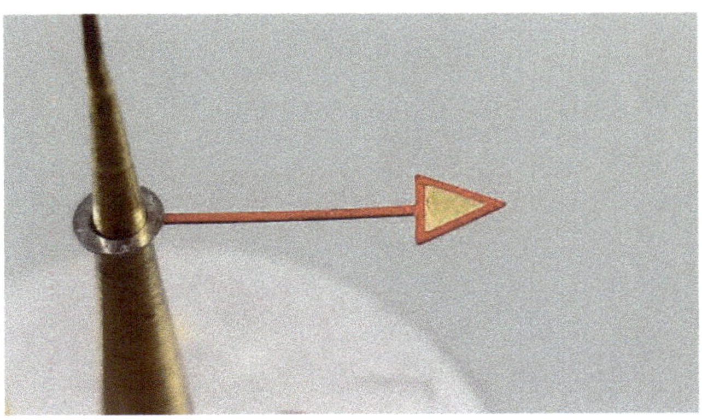

There remains controversy surrounding the authenticity of the all-red GMT hand

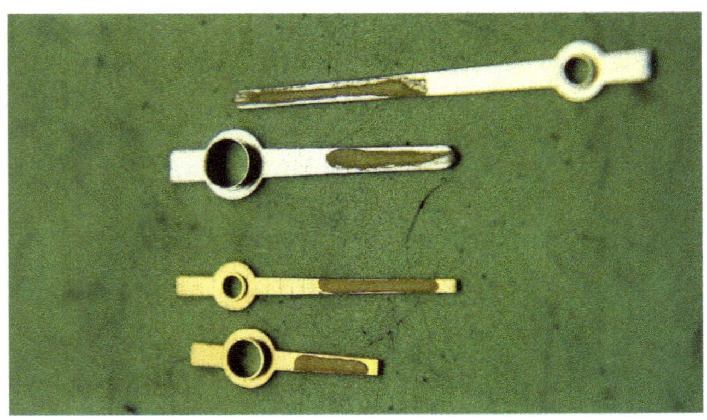

Lume strips on the underside of baten hands

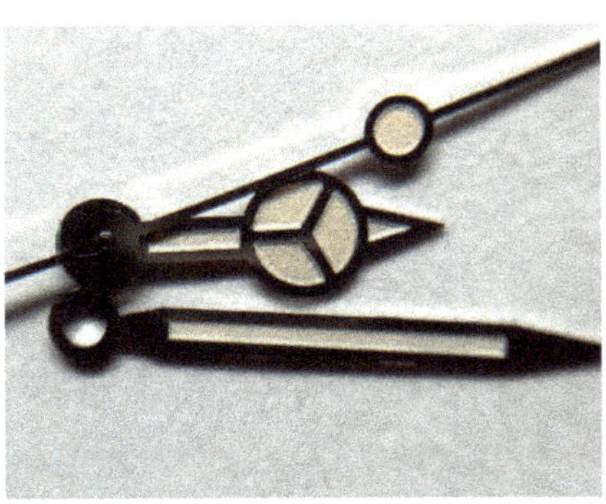

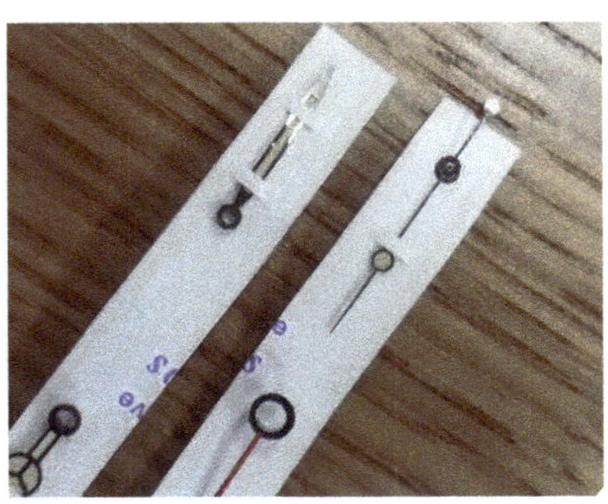

Vintage hands may exhibit various stages of corrosion depending on whether they are white gold, or electroplated

WINDING CROWNS

Rolex has patented several crown designs. Most were supplied by Boninchi SA, which was eventually acquired by Rolex in 2001. A wider variety of styles were used in the antique genre than during the vintage and classic genres. Later watches use one of only two crown styles.

The basic, or simple, crown style is used on Cellini and dress watches. This is a conventional pull-out style, typically with a raised coronet on the exterior surface. An onion-style is often seen on the antique models with cushion cases.

A screw-down crown is used on Oyster cases, with one exception. The non-screw-down Super Oyster crown was advertised as water resistant and was used on early Oyster references (such as the Oyster Precision and Oysterdate) in the 1950s. These are increasingly rare, and many were replaced during servicing. Early screw-down crowns may feature a Swiss cross and the word Brevet, meaning 'patented'.

There are two varieties of screw-down crown: Twinlock and Triplock. The Twinlock is recognized by either no marking beneath the coronet or a single line. These are used on classic crossover style watches like the Datejust and Oyster Date, etc. Early versions may also feature the word Brevet, as well as a Swiss cross.

The Triplock has three small raised dots below the coronet. These crowns are used on Professional models like the Daytona, Submariner, Sea-Dweller, GMT-Master, etc.

It is common to see aftermarket crowns or incorrect crowns on vintage Rolex watches. This obviously affects authenticity and value.

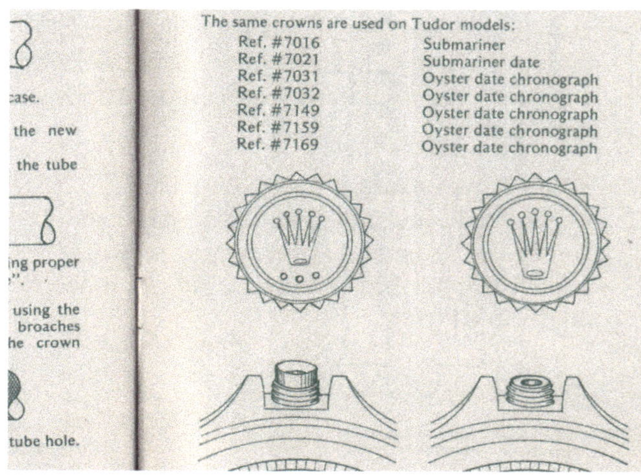

Extract from a technical service bulletin circa 1960s

Twinlock in Steel or Two-Tone, Gold and Platinum

Triplock in Steel or Two-Tone, Gold and Platinum

CRYSTALS

Rolex crystals are either acrylic or synthetic sapphire. Acrylic was cutting edge technology during the dawn of plastics, as was synthetic sapphire when introduced.

Acrylic crystals without the Cyclops date magnifier are called Tropic crystals.

Crystals are consumable components and were intended to be replaced at service intervals. An original crystal is somewhat meaningless and doesn't command any collector premium (except for pieces claiming to be NOS).

Acrylic crystals can be polished, unlike sapphire. Collectors use a variety of techniques and materials to buff out scratches and scuffs. Two of the most popular things used as polish are toothpaste and Polywatch.

Sapphire crystals on modern classic models may feature a laser-etched coronet (LEC) at the 6 o'clock position. This is an anti-counterfeiting measure and will require a loupe or excellent eyesight and lighting to see.

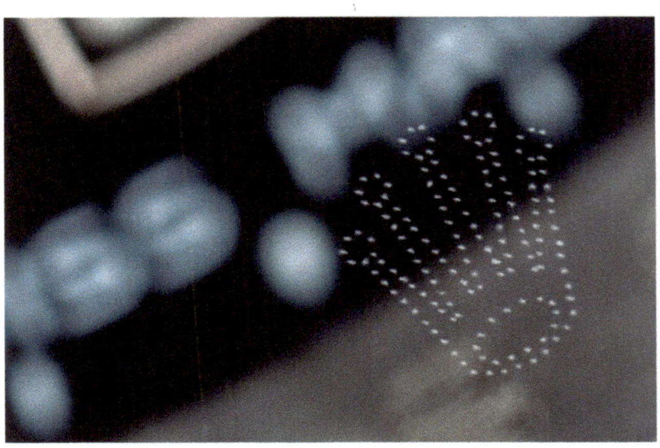

LECs were introduced in 2001 and had spread throughout Rolex's range by 2003, with the notable exception of the Milgauss Anniversary GV Edition. This unique green-tinted crystal (Glace Verte) does not have a LEC.

A high-domed, acrylic crystal has a distortion effect much beloved by vintage collectors

BEZEL INSERTS

A bezel insert is the small disc of aluminum that is pressure fitted into a bezel ring. They are used on Submariners, Sea-Dwellers, GMT-Masters, and early Turn-O-Graphs.

The numerals that appear on the inserts are created through a galvanic process similar to that used for gilt dials. They are not surface-printed. Therefore, numerals can not be scratched off. Rather, surrounding paint may be damaged, revealing the underlying color coat.

Over the decades, bezel inserts have been produced by several subcontractors, giving rise to many minor differences in font style, font size, and font positioning.

These bezel variants are the subject of many forum threads, lively and passionate discussion, and much heated, and strongly-held opinion. Despite this, there are few hard and fast rules governing which bezel styles are correct for which models and which serial number ranges. The quest for absolute correctness is a foolhardy one and fraught with exceptions.

There is a thriving market for these small metal rings, and counterfeits are abundant. Provided the bezel insert is authentic Rolex, cosmetically consistent with the dial and fits correctly into the bezel ring, the collector should be happy. The question of whether it's a service insert or the wrong MK for the serial range should not be a deterrent in buying a watch you like.

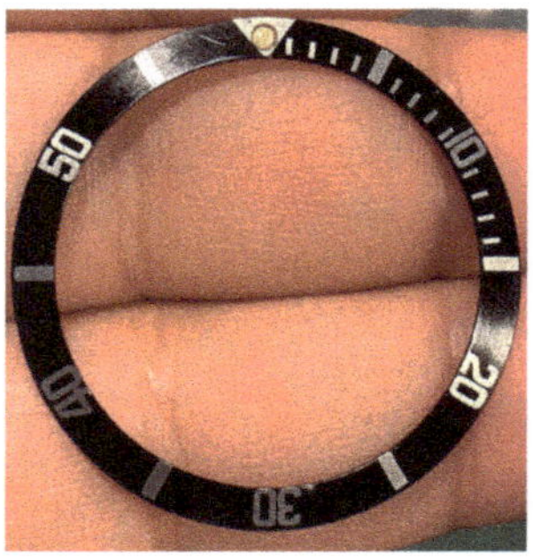

Submariner MK I

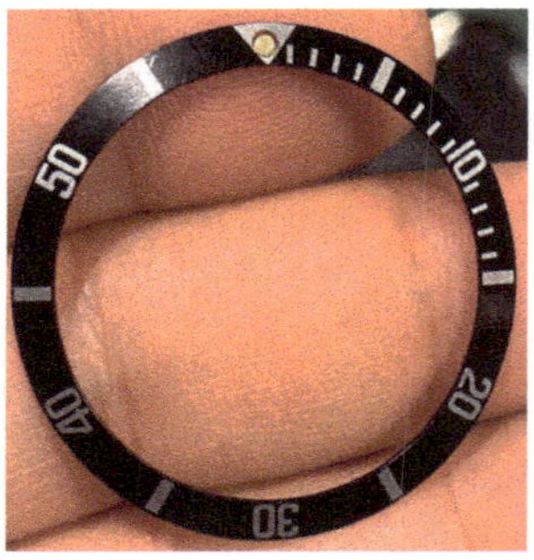

Submariner MK II

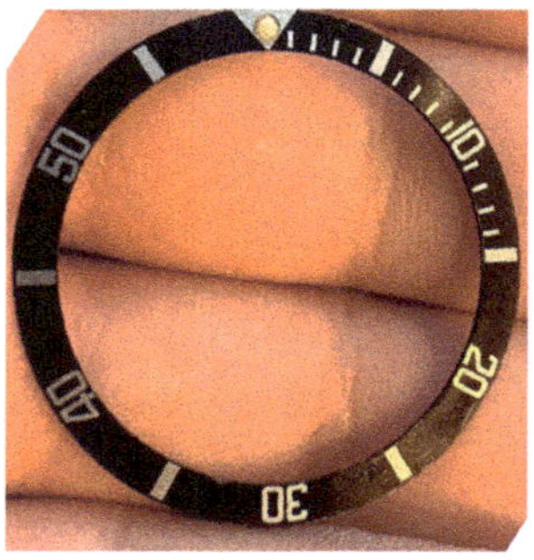

Submariner MK III

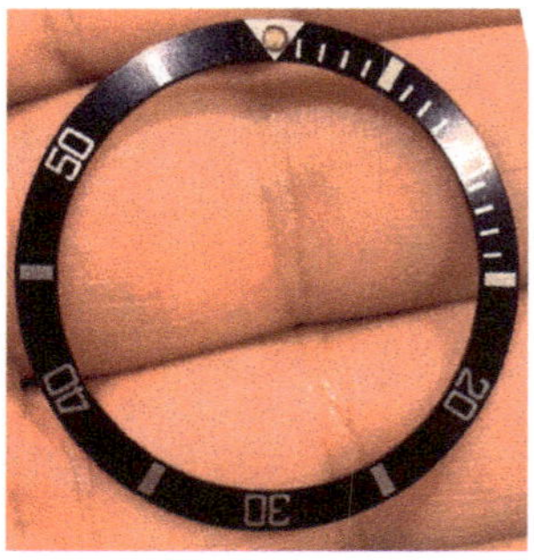

Submariner MK IV

Submariner unclassified late 1970s to early 1980s

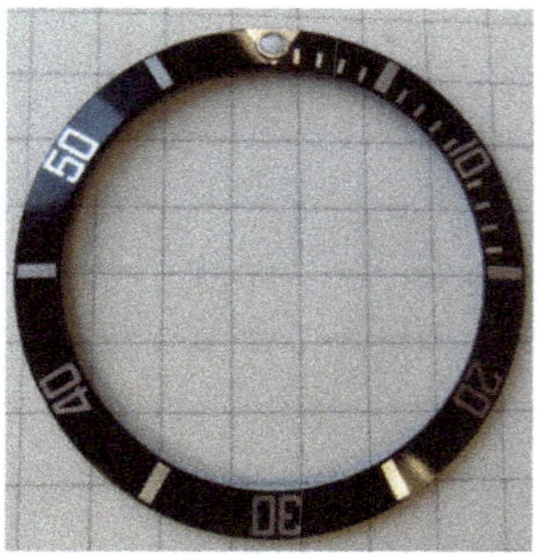

Submariner service insert 1990s

Late 70s early 80s Bezel

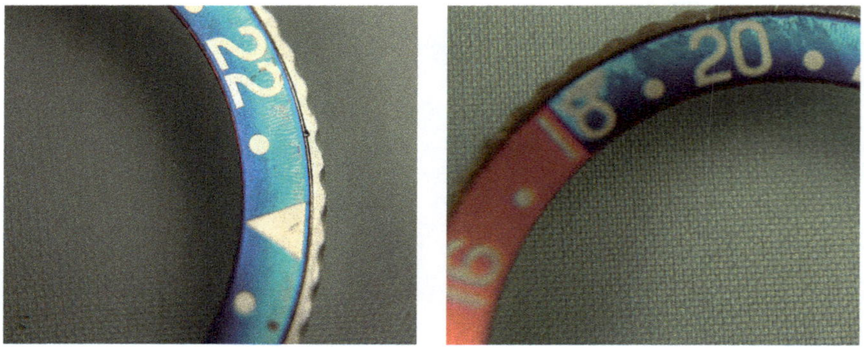

The blue pigment deteriorating to reveal the underlying silver color base layer

Chapter

06

ANTIQUE & POCKET WATCHES

During the earliest years, Rolex experimented with brand names liberally and interchangeably. Names like Prince and Imperial were mixed and matched across watches and pocket watches. Many of these early watches do not have case reference numbers or movement serial numbers.

The First World War (1914 to 1918) saw Rolex introduce trench watches. These were the first attempts by Rolex at a robust, tool-like, wrist-worn timepiece. They retained many design elements of the pocket watches, which were sold alongside them, including hinged case covers.

Similarities included brittle but highly visible enamel dials, often with the 12 printed in red. This helped the wearer adjust to having their timepiece on their wrist, where the up position may appear oriented counterclockwise 90 degrees compared to pocket watches. In the chaotic, low light of trench warfare, these small usability details were crucial.

Today's market for Rolex trench watches is strong, but the condition of the enamel dials is highly variable. Seldom worn by collectors, the top condition examples are highly prized at auction and carefully stored away as investments and heirlooms.

Other curious watches include stopwatches, pendant watches, and portfolio watches, which folded into a purse-like travel case. Many of these folding-style timepieces were made in the 1930s and 1940s and came to be called purse watches.

Pocket watches were the first products from Rolex and dominated sales until the start of WWII. Catalogs from the 1930s show open-face, semi- and full hunter style pocket watch cases. They were made in a variety of materials, from rolled gold and silver to solid 18K gold. In 1932 Rolex launched the Rolex Prince Imperial pocket watch in 18K gold. It turned out to be one of the most accurate watches made by Rolex and was sold alongside the Prince wristwatch. Rolex continued to make some very exclusive luxury pocket and pendant watches, but they were soon eclipsed by the success of their wristwatches.

We associate Rolex pocket watches with the war-era antique period, but Rolex was still making contemporary pendant pocket watches into the 1980s. These precious metal curiosities are not especially collectible but undoubtedly rare and unusual. They command only a modest price premium over their base metal and jewel value.

Antique Rolex pocket watches are more typical of the 1910s and 1920s, and appear in solid yellow or rose gold. A typical example would be the 44mm ref. 1528 with a pop-open caseback. By the 1930s, Rolex had retreated from this line of business, focusing on wristlets for both men and women.

> "I'm very proud of my gold pocket watch. My grandfather, on his deathbed, sold me this watch."
> *Woody Allen*

An antique 1930s pocket watch in 14K gold featuring an enamel dial. Sold in Australia and awarded to Mr V.H. Carkeek by his employer, J.Kitchen & Sons. in recognition of his long service.

POCKET & FOLIO REFERENCES

Class	Name	Ref	Cal	Case	Description
Purse	Sporting Prince	1561	Extra Prima	SS	Purse Watch
Purse	Prince Imperial	1585	Class A Observatory	SS	Travel Watch, enamel dial with sub-seconds (1940s)
Purse	Sporting Prince	1599	Extra Prima	YG	Purse Watch, guilloché case (1930s)
Pocket	Prince Imperial	1645	5096	YG	Pendantwatch 42mm, Extra Prima Chronometer (1940s)
Pocket	Prince Imperial	2939	Chronometer	WG	Pocketwatch, sub-seconds, 42mm
Pocket	Cellini	3717	1600	YGWG	Pendantwatch, Round, Ladies (1960s)
Pocket	Cellini	3727	1600	YGWG	Pendantwatch, Rectangular, Ladies (1970s)
Pocket	Cellini	3729	1600	YGWG	Pendantwatch, Oval, Ladies (1970s)
Pocket	Cellini	3790	1600	WG	Pocketwatch, Round
Pocket	Cellini	3791	1600	WG	Pocketwatch, Rectangular
Pocket	Cellini	3799	1600	WG	Oval

SS : Stainless Steel | YG : Yellow Gold | WG : White Gold

ROLEX PRINCE

When launched in 1928, the Prince received rave reviews. Customers loved it, declaring it avant-garde and a la mode.

The rectangular case had a large top dial for the hours and the minutes and a lower subdial for the seconds. Both the dial text and case embellishments were distinctly art deco. Designed as a luxury dress watch, it was an immediate hit, and the Prince and its derivatives remained in production for nearly three decades.

The Rolex Prince models were the first Rolex watches made in large quantities to achieve Chronometer Certification. It is now recognized as the commercial breakthrough that put Rolex on the map.

The subdial, with its jumping seconds indicator, became popular with doctors as a convenient aid in measuring a patient's pulse rate. The style subsequently came to be known as The Doctor's Watch.

The Prince was offered in two case styles – a more rectangular Classic style (ref. 1343) and a more round design known as Brancard style (ref. 971). Several materials were used to manufacture Princes, including yellow gold, sterling silver, rose gold, platinum, two-tone combinations and later, steel. Soon after introduction, the manual-wound movement was replaced by an automatic one.

In 1935, Rolex launched the Railway Prince (ref. 1527). The design was inspired by locomotives and today acts as a reminder of the luxury of European rail travel in the 1930s.

The Railway Prince introduced a unique Jumping Hours complication, which was a genuinely new and futuristic design. The upper dial had only a minute hand and the hours could be read as numbers (1-12) through a window at the 12 o'clock position. Years later this provided the inspiration for the Rolex Datejust.

Of the Rolex Prince series, the last was the Super Precision Aerodynamic. This had one big dial for the hour, minute, and second hands.

Sometime during the late 1940s, production of the Rolex Prince came to an end and the line was quietly retired.

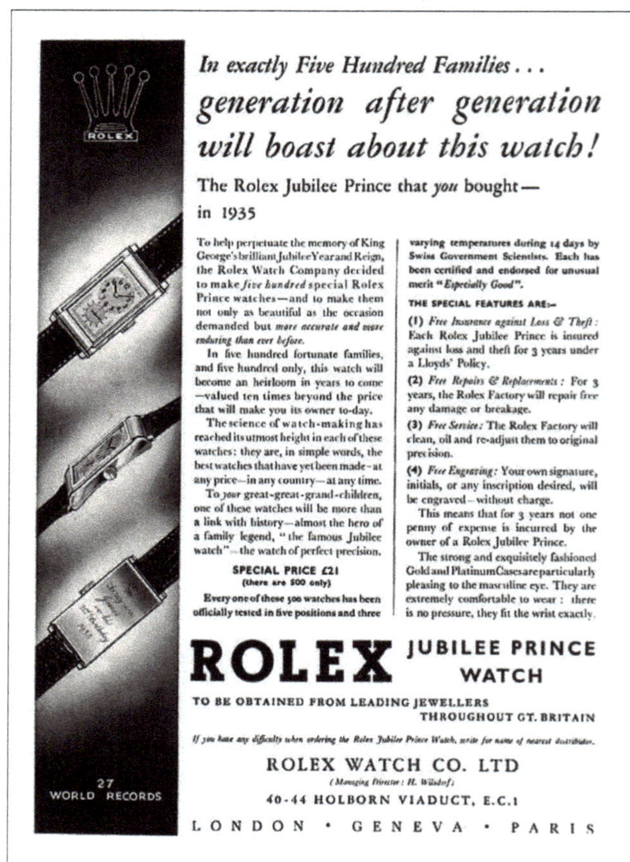

SPECIAL EDITION PRINCES

Several special editions of the Rolex Prince were made over the decades. The first was in 1935 when Rolex offered a special limited-edition Rolex Jubilee Prince. This short run of (a claimed) 500 units celebrated the Silver Jubilee of King George V.

The Sporting Prince was another limited edition created to test the market. It was an unusual pocket watch, designed to be used during athletic activities. The movement had a spring mechanism with which the dial could pop up from the case.

A well-known special edition Prince was the Quarter Century Club. These were made for a Canadian retailer to be gifted to employees who had achieved 25 years of service. These watches had the 12 letters of 1/4 Century Club printed around the dial instead of the hour or minute markers.

ROLEX PRINCESSE & ROLEX QUEEN

In 1930, Rolex launched the Rolex Princesse, a Ladies version of the Prince line. These were simple two-handed affairs (hour, minute and no seconds hand) offered in rectangular cases that were similar to, but smaller than, the Prince.

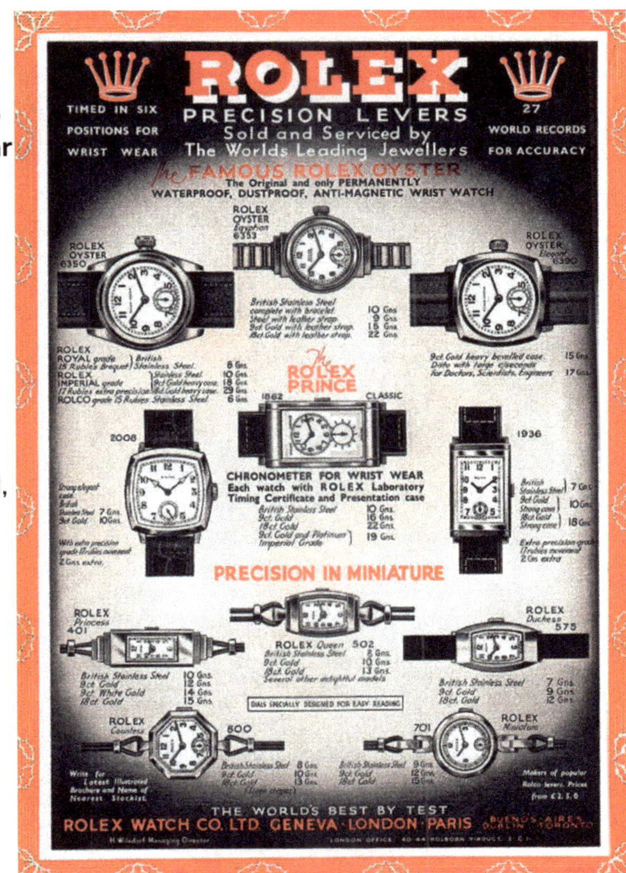

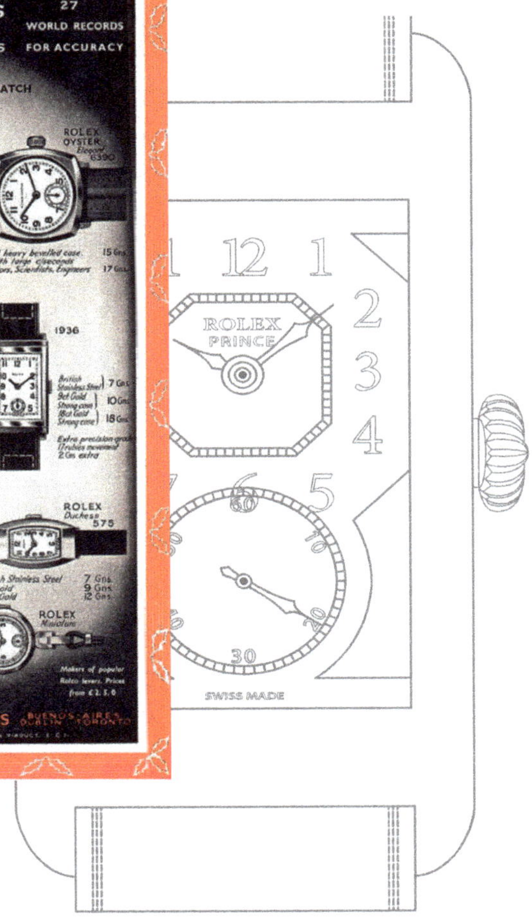

The Princesse was followed in 1932 by the Rolex Queen collection (refs. 503, 504 and 505). These three different rectangular case designs were offered in 9K, 18K or stainless steel.

The market for these editions is small, attracting little collector interest. Unlike the larger Prince models, they are largely neglected, and examples can be found in surprisingly original and unrestored condition. Women's models, in general, are not sought after. Even the few female collectors prefer the larger men's sizes. Ellen DeGeneres has an amazing vintage Rolex collection – all in large sizes.

Today, the Princess brand is a Tudor asset.

PRINCE, PRINCESS & QUEEN REFERENCES

Class	Name	Ref	Cal	Size	Case	Description
Wristlet	Prince Brancard	**971**	7½'''	Full Size	YG	Rectangular, sub-seconds
Wristlet	Prince Brancard	**1490**	300	Full Size	YG	Rectangular, 1/4 Century Club
Wristlet	Prince Brancard	**1491**	7½'''	Full Size	YG	Rectangular, Jumping hour, sub-seconds
Wristlet	Prince Brandard	**5068**	7½'''	Full Size		Rectangular, Jumping hour
Wristlet	Prince Chronometer	**1541**	7½'''	Full Size	YG	Rectangular, sub-seconds
Wristlet	Prince Chronometer	**3361**	7½'''	Full Size	YG	Rectangular, center-seconds
Purse	Prince Imperial	**1585**	Class A Observatory	Full Size	SS	Travel Watch, enamel dial with sub-seconds (1940s)
Pocket	Prince Imperial	**1645**	5096	Full Size	YG	Pendantwatch 42mm, Extra Prima Chronometer (1940s)
Pocket	Prince Imperial	**2939**	Chronometer	Full Size	WG	Pocketwatch, sub-seconds, 42mm
Wristlet	Prince Railway	**1527**	7½'''	Full Size	WGRG	Rectangular Stepped Case, sub-seconds
Wristlet	Prince Railway	**1862**	Extra Prima	Full Size	SSYG	Rectangular, sub-seconds
Wristlet	Prince Railway	**4376**	7½'''	Full Size	YG	Rectangular, sub-seconds
Wristlet	Princess	**1772**		Ladies	YG	Rectangular
Wristlet	Princess	**2405**		Ladies	SSYG	Rectangular
Wristlet	Princess	**5877**		Ladies	YG	Rectangular
Wristlet	Princess Egyptienne	**1382**	170	Ladies	RG	Rectangular
Wristlet	Princess Egyptienne	**1466**	170	Ladies	WG	Rectangular
Wristlet	Queen	**503**		Ladies		
Wristlet	Queen	**504**		Ladies		
Wristlet	Queen	**505**		Ladies		
Pre-Cellini	Queen Midas	**3581**		Ladies		Pentagonal
Purse	Sporting Prince	**1561**	Extra Prima	Full Size	SS	Purse Watch
Purse	Sporting Prince	**1599**	Extra Prima	Full Size	YG	Purse Watch, guilloché case (1930s)

SS : Stainless Steel | YG : Yellow Gold | WG : White Gold | RG : Rose Gold

IMPERMEABLE
27508
RETARD
AVANCE

OTHER ANTIQUE WRISTLET REFERENCES

Ref	Cal	Bezel	Size	Case	Description
13				YG	Rectangular
64			Ladies	WG	Rectangular
162				YG	Cushion
406			Ladies	RG	Tonneau
478				SS	Rectangular
514				YG	Rounded
556	Ultra Prima			YG	Rectangular
757				YG	Rectangular
758				YG	Rectangular
860				YG	Rectangular
870				YGWG	Rectangular
912	Prima			YG	Square
1027	Chronometer			RG	Round
1122	Prima			Silver	Round
1328				YG	Rectangular
1370				Silver	Tonneau
1383			Ladies	YG	Rectangular
1394			Ladies	Silver	Cushion
1401			Ladies	WG	Square
1618	Perpetual			GF	Round
1696	Extra Prima			WG	Rectangular
1832				YG	Rectangular
1852	Extra Prima			YG	Rectangular
1880	Ultra Prima			SSYG	Rectangular
1897				SS	Rectangular
1918				WG	Rectangular
1936				YG	Rectangular
1966			Ladies	SS	Tonneau
1992				SS	Rectangular
2007			Boys Size	YG	Cushion
2151				SSYG	Rectangular
2164			Ladies	WG	Round
2175			Ladies	YG	Square
2356				YG	Rectangular
2361	Chronometer		Ladies	YG	Rectangular
2372				YG	Rectangular
2524			Mid Size	SS	Round
2536				YG	Rectangular
2537	Ultra Prima			YG	Rectangular
2565	Ultra Prima		Boys Size	SS	Round
2568				SS	Round
2582	Ultra Prima			SSYG	Round
2694			Ladies	RG	Rectangular
2734				SS	Round
2886				SS	Rectangular
3003				SSRG	Round
3028				RG	Round
3029	Chronometer			RG	Round
3038				YG	Round
3059	Precision			RG	Rectangular
3073			Ladies	YG	Square
3078			Mid Size	SS	Round
3118			Ladies	YG	Octagonal
3140	Precision			YG	Rectangular
3251			Ladies	YG	Round
3255	Chronometer			YG	Square
3260	Chronometer			YG	Square
3265				SS	Round
3287				RG	Square
3428	Prima			YG	Cushion
3456				YG	Rectangular
3562			Ladies	SS	Square
3571				SS	Round
3573				YG	Round
3667	Precision			YG	Round
3684				SSYG	Square
3689	Chronometer			YG	Square
3737	Chronometer			YG	Square
3754				YG	Round
3762	Chronometer			YG	Round
3771			Ladies	YG	Rectangular
3777				YG	Rectangular
3782				RG	Round
3783	Chronometer			RG	Round
3861	Precision			SS	Rectangular
3861				YG	Rectangular
3894	Chronometer			YG	Square
3911			Ladies	RG	Square
3923				YG	Round
4029				YG	Rectangular
4051	Chronometer			SS	Round
4100				YG/WG	Horseshoe
4101				YG	Horseshoe
4102				WG	Horseshoe
4107				SS/RG	Round
4108				YG	Rectangular
4117	Chronometer			RG	Round
4118	Chronometer			RG	Round

Ref	Cal	Bezel	Size	Case	Description
4119				YG	Round
4134	Chronometer			RG	Round
4184			Ladies	SSYG	Rectangular
4211			Ladies	YG	Square
4289			Ladies	YG	Rectangular
4291			Ladies	YG	Square
4294			Ladies	YG	Square
4325			Full Size	YG	Round
4326				SS	Round
4330				YG	Rectangular
4334			Ladies	RG	Hunter
4364			Full Size	YG	Round
4366				SS	Square
4374	Precision		Ladies	YG	Round
4381			Ladies	YG	Square
4391			Ladies	YG	Square
4401			Ladies	YG	Square
4405			Ladies	YG	Square
4409	Chronometer			RG	Round
4411	Chronometer			YG	Round
4417				SS	Round
4446				RG	Round
4454			Ladies	YG	Square
4457			Ladies	YG	Round
4471				YG	Square
4473			Ladies	YG	Rectangular
4484			Ladies	YG	Hunter
4487			Ladies	YG	Round
4491			Ladies	RG	Round
4492			Ladies	YG	Round
4493			Ladies	YG	Square
4494			Ladies	YG	Square
4495			Ladies	YG	Square
4496	Precision		Ladies	RG	Round
4497	Chronometer	Fluted		YG	Rectangular
4513	Precision			SS	Rectangular
4533	Chronometer			YG	Rectangular
4542				SS	Round
4556			Ladies	SS	Round
4560				SS	Round
4579			Mid Size	RG	Round
4595	Precision		Ladies	RG	Rectangular
4612				WG	Oval
4613				WG	Oval

Ref	Cal	Bezel	Size	Case	Description
4615			Ladies	RG	
4643	Chronometer			YG	Square
4645	Perpetual		Ladies	YG	Square
4663	Perpetual		Mid Size	RG	Square
4725			Ladies	YG	Round
4816	Chronometer			YG	Round
4830			Ladies	SS	Rec
4845	Chronometer			YG	Round
5088			Ladies	SS	Rectangular
6512	Chronometer			YG	Round
7000				SSYG	Square
7008				YG	Round, Canadian Market
7011				YG	Round, Canadian Market
7038				YG	Round, Canadian Market
7051				YG	Square
8029	Precision		Full Size	SS	Round
8094				YG	Square
8126	Super Precision		Ladies	RG	Rectangular
8320			Ladies	RG	Square
8382				RG	Round
8522			Ladies	RG	Square
8569	Chronometer			YG	Round
8612				RG	Square
8651	Chronometer			YG	Round
8731	Precision		Ladies	WG	Round
8784			Ladies	RG	
8788	Precision		Ladies	YG	Round
8790	Precision		Ladies	RG	Square
8858				YG	
8952	Chronometer			YG	Round
9242			Ladies	RG	Square
9250			Ladies	RG	Rectangular
9347	Chronometer			YG	Rectangular
9576	650		Full Size	YG	Round
9578	650		Full Size	YG	Square
9638			Ladies	RG	Square
9720			Full Size	WG	Square
9798				WG	Round
9878	650		Ladies	RG	Square

SS : Stainless Steel | YG : Yellow Gold | WG : White Gold | RG : Rose Gold | GF : Gold Fill | LN : Lunette Noir

MOON PHASE & CHRONOGRAPHS

Early Rolex chronographs are not well documented. During this time between world wars, there was little media advertising of luxury goods, making Rolex catalogs and print advertisement scarce. Of those that were printed, very few survive. This lack of archival history makes verification and authentication of these early watches very difficult.

Rolex used the same Valjoux movements as other chronograph makers of the time. These watchmakers included Longines, Heuer, and Breitling, each going to varying lengths to customize their Valjoux movements. This practice is known as finishing.

Rolex finished their movements by signing the balance cock and the train gear bridge. They went to considerable effort to apply satin finishes and bevels to individual parts. Finishing styles evolved over time, and each manufacturer favored different techniques. It takes an expert to identify Rolex finishing from different decades. Only a few of these experts exist, and they commonly work for the premium auction houses (if not Rolex itself).

Nearly all antique Rolex chronographs in collections today have at least some parts that were not made by Rolex. This makes the question of authenticity, originality, and correctness moot. Their value lies in their visual appeal and running order.

DATO-COMPAX CHRONOGRAPHS

A class of hand-wound chronographs produced between the 1940s and 1960s was known as the Dato-Compax. These have acquired the nickname, Jean Claude Killy. These span the boundary between what's considered Antique and Vintage.

Jean Claude Killy was a world champion skier who dominated the downhill slalom and giant slalom events in the 1960s. He also enjoyed a career as a race car driver and screen actor. He became a Rolex brand ambassador in the 1970s and went on to join the Rolex board of directors.

Five references bare the nickname Jean Claude Killy - refs. 4768, 4767, 5036, 6036 and 6236. Strangely, Rolex produced these references a decade or more before his skiing achievements. The nickname has been applied by collectors retrospectively.

The name Dato-Compax consists of two intricate complications – the date (Dato) and a chronograph (Compax). The term Compax was first used by Universal and Zenith around 1936. It referred to the number of complications of a movement. Later it came to refer to the arrangement of the triple subdials.

The ref. 4768 was the shortest-lived of the Killy Chronographs. Experts claim Rolex made only 220 units. It is the only Data-Compax without an Oyster case and featured a snap-on caseback. The dial features only the text, "Rolex Chronographe."

In 1947, Rolex replaced it with ref. 4767. This was a significant upgrade with an Oyster case and screw-down crown. It now had the dial text, "Oyster Chronographe." This is the first triple calendar chronograph in an Oyster case. The ref. 4767 lasted only a year.

It was replaced with ref. 5036, in 1948, which ran until 1951. The ref. 6036 replaced it a few years later around 1955. In the late 1950s, the ref. 6236 arrived. It was to be the final Dato-Compax. It had a larger bezel, reminiscent of today's three-piece case construction.

The Dato-Compax arrived on the market shortly after the end of WWII. Rolex offered them in stainless steel, yellow gold, and rose gold. Austere market conditions meant demand was soft, but sales were insufficient to sustain the high production costs of such a complicated watch.

These are still the most complicated watches ever produced by Rolex. The small production volume means they are scarce. When they come to auction, they attract considerable attention and command headline prices.

DATO-COMPAX REFERENCES

Ref	Cal	Depth Rating	Start	End	Case	Description
4767	V72	50m	1947	1948	SSYG	Twin-pusher, triple register, Oyster case, triple calendar
4768	V72		1947	1948	SS	Twin-pusher, triple register, triple calendar, fancy lugs
4769	V72		1947	1948	YG	Twin-pusher, triple register, triple calendar
5034	72		1948	1951	YG	Twin-pusher, triple register, Oyster case
5036	720		1948	1951	YG	Twin-pusher, triple register, triple calendar, Transitional
6036	72		1951	1955	SS	Twin-pusher, triple register, triple calendar
6234	72	50m	1955	1961	SS	Twin-pusher, triple register, Oyster case, screw down crown Tachymeter and telemeter on dial
6236	72C		1958	1962	SS	Twin-pusher, triple register, Oyster case, triple calendar

SS : Stainless Steel | YG : Yellow Gold

Jean Claude Killy, 1968. Triple Olympic Gold Medalist (Center)

CENTERGRAPH & ZEROGRAPHE CHRONOGRAPHS

Name	Reference	Caliber
Zerographe / Centregraph	**3346**	10 ligne Hunter
Zerographe / Centregraph	**3462**	10 ligne Hunter

These two references (3346, 3462) were the first Oyster-cased, monopusher chronographs. They are 32mm in diameter and were the first of any chronograph to use the Rolex Oyster crown. They are also the first Rolex watches to feature a rotating bezel.

They were housed in a Bubbleback case and were an entirely in-house chronograph, using a 10 ligne Hunter movement, which was retrofitted with a flyback chronograph complication. This was achieved by adding only ten additional parts to a conventional hunter movement, plus a protruding caseback to accommodate them.

Launched in the mid-1930s, they were made in small numbers and for only a few years. Little is known about them, since only a few have surfaced at auction (Phillips and Christie's) with little to no documentation.

At launch, the names Zerographe and Centergraph were used interchangeably on both references. As production settled, Zerographe was applied to ref. 3346 and Centergraph to ref. 3462.

The Zerographe (3346) has a rotating steel bezel used to measure elapsed minutes. A rotating bezel did not appear again until the Turn-O-Graph 20 years later in 1953. The Centergraph (3462) has a fixed polished bezel with a tachymeter on the edge of the dial.

Their dials appear in both black and white, with some examples in the California style (half Roman and half Arabic numerals).

These watches are extremely rare. To this day, they are the only flyback monopusher chronograph ever made by Rolex. Only one other reference (a Bubbleback) has used a California dial.

REF. 6062 MOONPHASE

This triple calendar moonphase model offered day, date, month and phase of the moon. It debuted at the Basel Watch Fair in 1950. Early advertisements describe it as a Cosmograph, yet this text does not appear on any of the dials.

An example of ref. 6062 sold at auction in May 2017 for CHF 5M, setting a world record for the most expensive Rolex watch. This record-breaking example featured a Bao Dai dial. Named after Vietnam's last emperor, it features diamond numerals.

There are only three documented examples of ref. 6062 with diamond dials. The Bao Dai is the only known example with the diamonds positioned on even numerals. Unlike the others, it has a diamond at the 12 o'clock position, displacing the Rolex coronet. The coronet appears below the diamond in place of the upper dial text. The traditional Rolex Oyster Perpetual text appears below the day and month apertures.

For the 1950s, ref. 6062 was a masculine 36mm. It used a 9 ligne movement and was the first watch with a water-resistant, non-screw-down, Super Oyster crown. These predated the screw-down Twinlock and were often replaced during service.

Sold alongside the Dato-Compax, ref. 6062 appeared in stainless steel, yellow gold, and pink gold. Dials were available in black and white with unusual combinations of hour indexes. One of the most eye-catching is the Stelline dial, which is Italian for "starlet dial". It had eight five-pointed gold stars as hour markers and a blue or black moonphase subdial. Other configurations include combinations of stars, arrowheads, and diamonds.

REF. 8171 PADELLONE MOONPHASE

In 1949, Rolex introduced the ref. 8171 which acquired the nickname "Padellone," Italian for "large frying pan." This was a triple date moonphase watch made in steel and yellow gold only. It had a snap back case, engraved with a serial number and coronet. Scholars believe Rolex made less than 1000 examples of ref. 8171.

By standards of the time, it was a substantial 38mm. It had sharply curved lugs protruding from the sides of the case ,emphasizing its size.

The Padellone featured an unusual cal. A295 automatic movement, which was not widely used in other references. The dial has twin apertures displaying the day and month. An integrated subdial shows both a fan-shaped moonphase and a running seconds hand. A fourth hand points to the date on an outer chapter ring.

This is a gorgeous and well-balanced watch. At auction, it changes hands for serious money.

OTHER ANTIQUE CHRONOGRAPHS REFERENCE

ANTIQUE CHRONOGRAPHS

Ref	Cal	Start	End	Description
1074	1560			Mono-pusher, twin register, cushion case
1223	1560			Mono-pusher, twin register, cushion case, telemeter
2021	16 1/2			Twin-pusher, twin register, tachymeter and telemeter
2022	V23			Mono-pusher, twin register, cushion case
2023	V23			Mono-pusher, twin register, round case
2057				Mono-pusher, twin register, cushion case
2226				Mono-pusher, twin register, boys size round case
2303	10 1/2			Mono-pusher, twin register, boys size round case
2507				Mono-pusher, twin register, round case
2508	V23			Twin-pusher, twin register, tachymeter and telemeter
2705				Twin-pusher, twin register, tachymeter and telemeter
2737				Twin-pusher, twin register at 12 and 6
2811	72A			Twin-pusher (round or oval), twin register, boys size round case
2917				Twin-pusher (square), twin register
2918	V23			Twin-pusher, twin register
2920	V23			Twin-pusher, twin register
3036				Twin-pusher (round), twin register
3055	69			Twin-pusher (oval), twin register, full and boys size round case
3082	V72			Twin-pusher (oval), twin register
3085				
3181				Twin-pusher (round), engine-turned bezel
3233				Twin-pusher, twin register, tachymeter and telemeter
3330	V22			Twin-pusher, twin register, tachymeter
3333	V22			
3335	V22			Twin-pusher (square), triple register, pulsemeter
3481				Twin-pusher (round) twin register, mid-size round case
3484				Twin-pusher, twin register
3525	12			Twin-pusher, twin register
3529	10 1/2			Twin-pusher, twin register, square case
3635				Twin-pusher, twin register
3642				Mono-pusher, twin register, small round case
3668	13			Twin-pusher, twin register, engine turned bezel, screwdown crown
3695				Twin-pusher, twin register

Ref	Cal	Start	End	Description
3735				Mono-pusher, twin register, enamel dial, hinged lugs and caseback
3827				Twin-pusher, triple register
3834	10			Twin-pusher, twin register, tachymeter and telemeter
3835				Twin-pusher, triple register, tachymeter and telemeter
3997				Twin-pusher, twin register
4048	V23			
4062	V23			Twin-pusher, twin register
4099				Twin-pusher, twin register
4100	13			Twin-pusher, twin register
4113	V55	1942	1942	Twin-pusher, twin register, split seconds, large case, Official Timer for Formula 1
4311				Twin-pusher, twin register
4313	V23			Twin-pusher, triple register
4332	V27			
4352				Twin-pusher (round), twin register
4500	V23			Twin-pusher, twin register, screw down crown
4537	V72			Twin-pusher, triple register
6034	72			Twin-pusher, triple register, screw down crown
6232	72			Twin-pusher, twin register
6238	72B			Twin-pusher, triple register, polished bezel, screw down crown
6270	55			
7131	234			
7132	234			
7159	234			
7169	234			
8206	69			Twin-pusher, twin register, square case
9162	V72	1954		Twin-pusher, triple register
6032	72			Twin-pusher, twin register, triple calendar
6236	72B			Twin-pusher, triple register, triple calendar
8180	172			Twin-pusher, triple register, triple calendar
8237				Twin-pusher, twin register, triple calendar, moon phase
6062	780	1949	1953	Twin-pusher, triple register, triple calendar, moon phase
8171	A295	1949	1952	Perpetual Chronometer, triple calendar, moon phase, Padellone

Padellone | pa.del'lo.ne |

noun ***Latin***

- Augmentative of ***padella***.
- a large frying pan or skillet.
- a bedpan.
- a most unique and rare Rolex triple calendar reference 8171

GENEVA R SWISS

ROLEX

PATENTED

1949-1954

Chapter 07

PROFESSIONAL TOOLS

In Rolex marketing terminology, these watches belong to the Professional Collection. Rolex adopted the name Professional for their collection of sport and tool watches around the year 2000. The collector community still refers to them as Tool and Sport watches, and dislikes the similarity to Omega's Professional Collection (Speedmaster, Seamaster, Railmaster).

While there are only eight model lines in the Professional collection, there are several versions of them within it. Each of these has many small variations in seemingly insignificant details.

There are approximately 196 distinct references to consider since the Rolex tool watch was first introduced in 1952. This figure excludes semi-tool watches like the Turn-O-Graph and Killy Chronographs, which are covered in other sections.

"First he wrought, and afterward he taught."
Geoffrey Chaucer

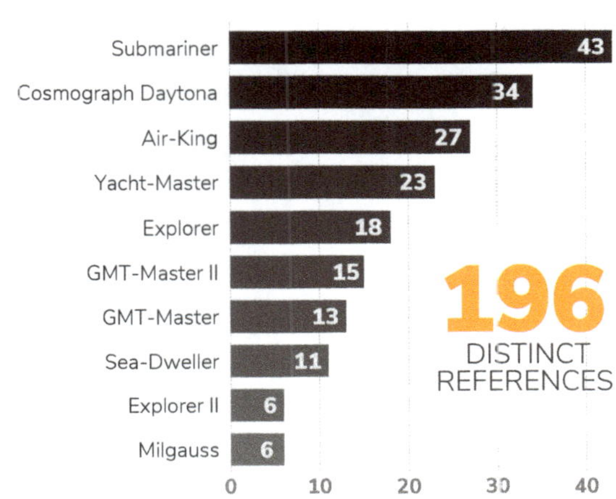

ROLEX IN AVIATION AND SPACEFLIGHT

If you're interested in acquiring a vintage tool Rolex, you will quickly learn about the nautical, aviation, and mountaineering history. The role played by Rolex in early spaceflight is often overshadowed by Omega and the iconic Speedmaster. From early supersonic test flights to the last of the NASA space shuttles, a Rolex wristwatch has been present. Favored by Soviet cosmonauts too, Rolex watches have served time on the International Space Station and other orbital missions.

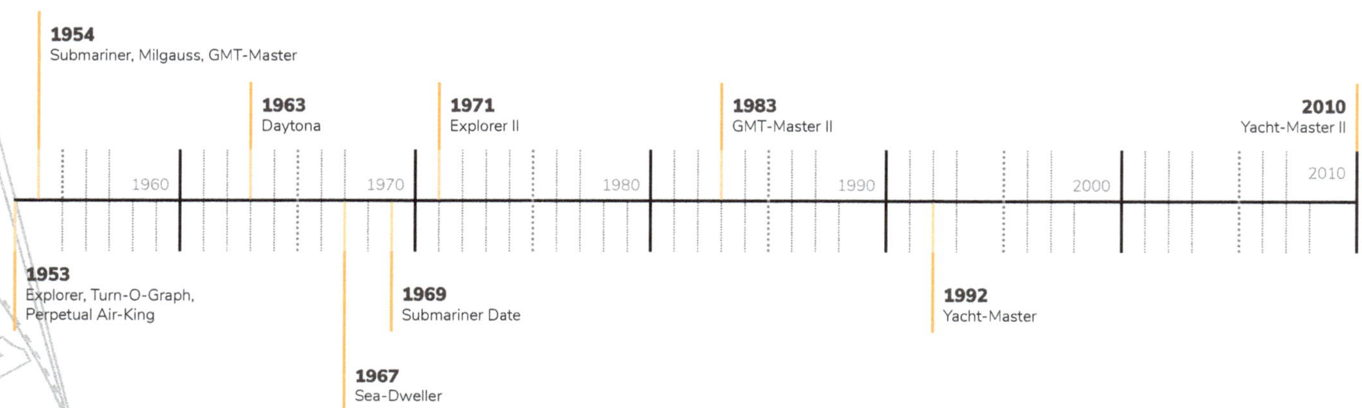

A Quick History Recap

In 1908 at age 28 and after a decade working in Swiss watchmaking, German-born Hans Wilsdorf (1882-1960) founded Rolex watches. The name Rolex was short, hard to misspell and easily pronounceable in many languages.

In 1914, at the outbreak of World War I, Wilsdorf foresaw the rise of wristwatches and the eventual demise of pocket watches. These 'wristlets' were to later earn their reputation as military tools in the bloody trenches of war. Trench-watches were no longer feminine trinkets but essential equipment for artillery officers. Rolex quickly acquired bona fide tool-watch credentials.

At about the same time, Rolex became the first wristwatch in history to be awarded a Class A certificate from the Kew observatory in London. A few years later in 1927, the Rolex Oyster was patented and launched as the first commercially viable waterproof wristwatch. By this time, Wilsdorf had focused engineering efforts on hardening and toughening his watches to withstand the rigors of the harshest environments.

By October 1940, Rolex had further developed the Oyster into the Air-series wristwatches to honor the bravery of Royal Air Force pilots during the Battle of Britain. The white-cream dial Air-Giant, Air-Lion, and Air-Tiger almost immediately became favorites with military pilots. Since 1945, the Air-King has embodied the aviation heritage of the original Royal Oyster.

By the end of World War II, the evolution of the Oyster case had come to include a screw-down crown. These innovations were combined with the new Perpetual rotary winding mechanism, and together with creative advertising, Rolex became the world's most well-known luxury brand.

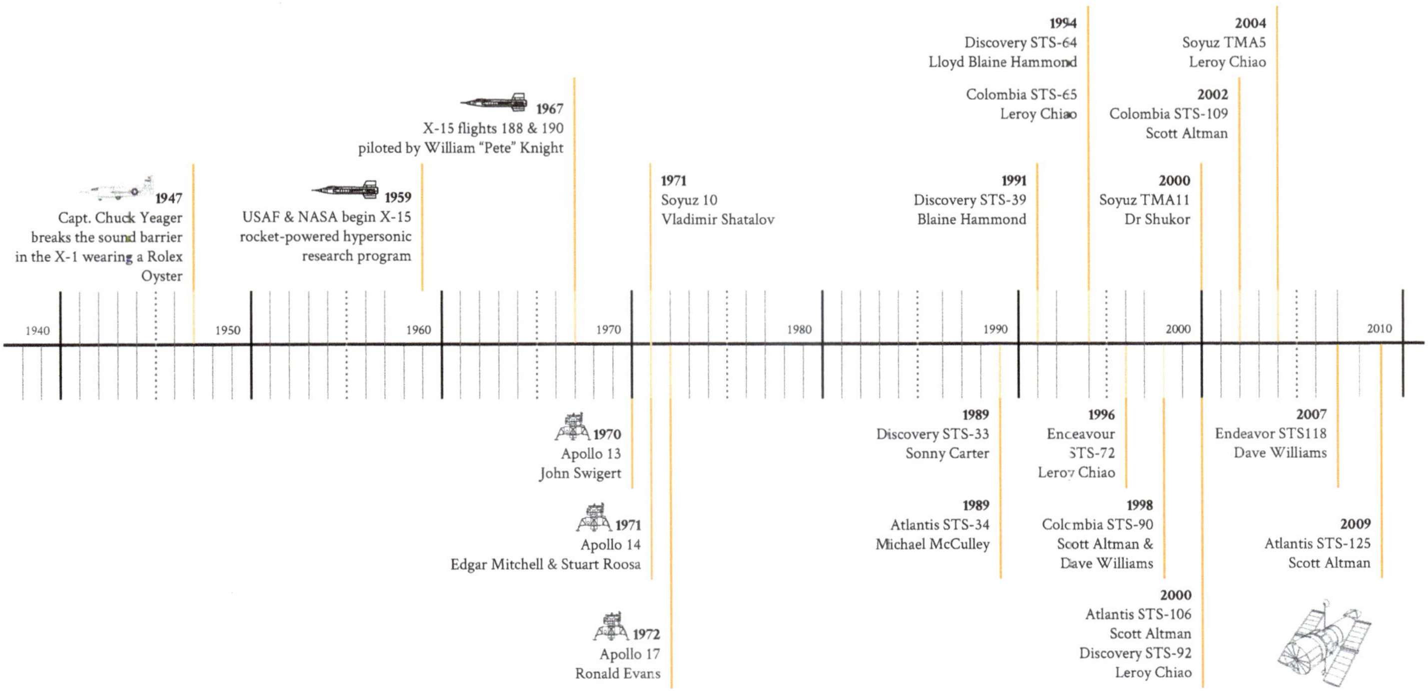

Rolex watches have been space-flown over 50 years, worn by 15 astronauts and cosmonauts, aboard 20 missions.
Rolex watches functioned in zero gravity for a collective 515 days covering well over 53 million orbital miles.

The story begins on October 14th 1947, a white cream dial Rolex Oyster literally skyrocketed on the wrist of Charles "Chuck" Yeager as he became the first pilot to break the sound barrier in level flight. During the rest of his career, Yeager turned out to be a die-hard fan of Rolex watches as he wore consecutive models such as a Submariner 6538 (1958), a GMT-master 6542 (1962) and a GMT-master 1675 (1971).

By 1950, Rolex had perfected some game-changing firsts in the world of horology, such as a reliable automatic movement with a date complication. Rolex watches were appearing on the wrists of world leaders such as Sir Winston Churchill and US President Dwight Eisenhower. Wilsdorf's strategy of pairing top-quality product development with rugged use by joint military services and scientific expeditions, burnished an increasingly prestigious reputation. The accuracy and durability of Rolex led to the tool watch concept aimed at specific professions like engineers, divers, explorers, pilots, and astronauts.

In 1953, as commercial jet aviation matured, Rolex was asked by several airlines to design a wristwatch to fit the needs of aircrew and world travelers. Pan American Airlines and Rolex teamed up to develop a suitable timepiece that could simultaneously display time in two different time zones.

It all begins with the Bulova Accutron

By 1961, all X-15 test pilots and the seven Mercury astronauts were given a Bulova Accutron tuning fork watch, which was heralded as being designed from the ground up for use in the harsh cockpit environments of experimental aircraft and space capsules. In 1962, Bulova introduced the world's longest watch strap, 44 cm long made of nylon webbing similar to seat belt material, it allowed the Accutron to be worn over the forearm of the test pilot's pressure suit. To reinforce this fact, the word "Astronaut" was added to the dial of the Bulova Accutron watches.

An acknowledged fake white-dial GMT-Master 6542

The white dial GMT-Master

There is a GMT-Master 6542 with a distinctive white dial. Very few have surfaced and they are controversial. Urban legend tells of the white dial being issued to PanAm ground crews and executives, with the black dial reserved for flight crews. This story has not been substantiated and the authenticity and legitimacy of white dial GMT-Masters is hotly debated.

This is as controversial as the GMT 'Blueberry', a 1675, with an all-blue aluminum bezel. Claimed to have been issued to Middle Eastern military leaders, their provenance and authenticity is hotly debated.

Both white dial and the Blueberry GMTs command high prices and should be considered high-risk acquisitions that have the potential to lose significant value in the future.

In 1954, Rolex launched the GMT-Master wristwatch. It featured a red and blue 24-hour bezel, a black dial, a date function, and a fourth 24-hour hand. This was differentiated by a distinctive luminous triangle at its tip. This Greenwich Mean Time (GMT) hand enabled flight crews to set the watch to another time zone by using the rotatable 24-hour bezel, and setting it to the correct offset.

The Rolex GMT-master wristwatch was quickly adopted by many airlines as the GMT standard was required for all aviation planning and navigation. The 1960s Rolex catalog heavily publicized the fact that both pilots and navigators considered the watch an important professional aid.

That same year the GMT-Master 1675 model became available, featuring protective crown guards. It was built for the purpose of timekeeping for the cockpit crew. This popular red and blue Pepsi-bezel watch remained available through 2007, and caught the eye of both American astronauts and Russian cosmonauts along the way.

In 1959, the United States Air Force started operations of the X-15 rocket-powered aircraft in NASA's hypersonic research program. The goal was to acquire research data to guide aircraft and spacecraft design.

In 1963, as NASA started recruitment for the Group III astronauts, Mercury astronaut John Glenn's goodwill tour brought him to Japan. He was met with great public enthusiasm and the event was noticed by Rolex. The company had wanted to release a watch in tribute to the brave spacefarers of the time, as it had done to commemorate expeditions to the summit of Mount Everest and the Arctic.

In response to the Japanese fascination with astronauts and space-related science fiction, Rolex launched a small production run of the 'Space Dweller', which was a modified version of the Explorer model. This model remained a Japan-only special edition and was not released to Rolex's world market.

On June 1, 1965, after four months of rigorously testing three watch brands, NASA's Gemini & Flight Support Procurement Office adopted the Omega

1958, Cpt. Chuck Yeager wearing a Submariner ref. 6538

1967, William "Pete" Knight inspecting the X-15-A2 before setting the air speed record

1967, William "Pete" Knight after setting an air speed record of Mach 6.7

Speedmaster as its official flight-qualified timepiece. This model was to be christened "the Moonwatch" in July 1969. Although Rolex missed out on a historic advertising opportunity, their watches were to make it into both the American and Soviet-Russian space programs.

Astronauts and test pilots privately purchased Rolex wristwatches, demonstrating a strong preference for the GMT-Master reference 1675. By 1965, Mercury astronaut Scott Carpenter became an aquanaut, joining the US Navy Sealab II project to test underwater habitats and advance the science of deep-sea exploration. During that time, he wore a Rolex Submariner and later ,a Rolex Sea-Dweller, the first wristwatch designed for saturation diving.

In June 1967, test pilot William "Pete" Knight was pictured wearing a Rolex GMT-Master while inspecting the ablative white coating on the X-15-A2 rocket-powered research aircraft. On October 3, 1967 Knight piloted this particular aircraft at a speed of Mach 6.7 (7274 km/h), a record that remains unchallenged today. Pete Knight's next X-15 flight on October 17, 1967 brought him to an altitude of 85.5 kilometers, high enough to be awarded the USAF astronaut wings. Post-flight photos revealed that he was wearing his Rolex GMT-Master on both of these X-15 flights.

A MODERN SPACE ARCHIVIST

Philip Corneille, Croatia

Since 1971 Philip Corneille, a Fellow of the Royal Astronomical Society, has been collecting and archiving historical photographs of significant aviation and spaceflight events. A retired NATO officer, his focus is the wristwatches worn by pilots, astronauts and cosmonauts in official portraits and mission imagery.

Philip has contributed to many journals, articles and books including this one and maintains a highly acclaimed photoblog at MoonwatchUniverse.tumblr.com. This curated archive is an invaluable resource and represents one of the most comprehensive archival records of space flown watches.

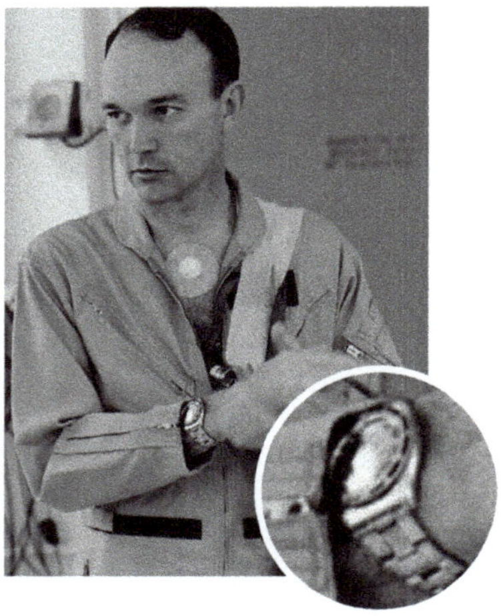

Michael Collins wearing an early Turn-O-Graph

ROLEX AND THE APOLLO MOON PROGRAM

In 1968 and 1969, several Apollo program astronauts were seen wearing Rolex wristwatches during training activities. These included Walter Schirra, Frank Borman, James Lovell and Michael Collins. However, the first Rolex flown aboard an Apollo mission was worn by Apollo 13 Command Module Pilot John "Jack" Swigert.

APOLLO 13

Just 72 hours prior to the Apollo 13 launch, Swigert replaced Ken Mattingly and was photographed wearing his personal Rolex GMT-Master while discussing details with Deke Slayton. April 11, 1971 Apollo 13 suit-up photos show Swigert wearing his personal Rolex with a bracelet on his left wrist underneath his spacesuit.

On April 13, 1970 Swigert spoke the now famous words, "Houston we've had a problem". Three days later on April 16, he used his Speedmaster to time the lunar module DPS (descent propulsion system) engine burn at 04:31:28 to :42.

Upon their return to Earth, Swigert gifted his space flown Rolex to Rolex Public Relations Director, Rene-Paul Jeanneret, who offered Swigert a gold black bezel Rolex GMT-Master on a jubilee bracelet. In December 1971, Jack Swigert finally posed for his official white spacesuit portrait wearing this gold watch.

John "Jack" Swigert wearing a gold GMT-Master, a gift from Rolex executive Rene-Paul Jeanneret

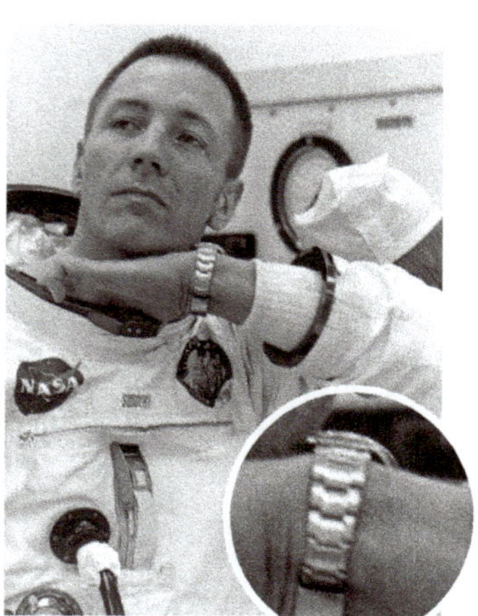

1970 John "Jack" Swigert suiting up for the Apollo 13 mission

Jack Swigert (left) with his GMT-Master at the Apollo 13 return ceremony Hickham Air Force Base near Honolulu. April 18, 1970

APOLLO 14

Between August 1969 and November 1970, the Apollo 14 crewmembers were pictured wearing Rolex GMT-Master watches during geology training as they prepared to bring moon rocks to Earth. NASA astronauts were allowed to carry private items in their personal preference kit (PPK), along with other small personal effects including wristwatches. Some of these fireproof-PPK bags were stored in the Command & Service Module (CSM), while others hung in the Lunar Module (LM). The latter descended to the lunar surface, rendering the items LM-flown.

During suit-up procedures on January 31, 1971, both Command Module Pilot Stuart Roosa, and Lunar Module Pilot Edgar Mitchell, were seen wearing Rolex GMT-Master watches. Moreover, Ed Mitchell had a pair of Rolex GMT-Master 1675 watches, tucking one on each wrist under the inner cuff of his A7L spacesuit. Although both Roosa and Mitchell clearly wore the Rolex in the CSM "Kitty Hawk", there's no footage showing Mitchell wearing the watches aboard the lunar module "Antares", nor on the lunar surface.

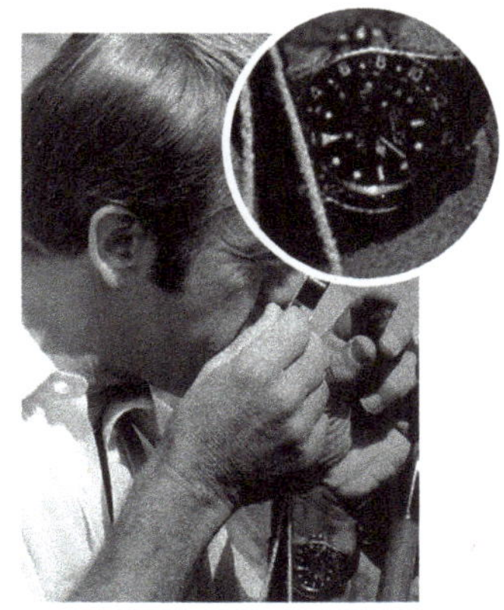

Alan Shepard and his personal GMT-Master during geology field trip

APOLLO 17

In 1972 during the November 21 lunar EVA-training at Cape Canaveral, Apollo 17, Commander Eugene Cernan was pictured wearing a Rolex GMT-Master strapped over the left forearm of his A7LB space suit. A few months later in December, Apollo 17 Command Module Pilot Ronald Evans was the last to account for a space-flown Rolex watch during the Apollo program. During the mission, Evans asked his moonwalking colleagues, Gene Cernan and Harrison Schmitt, to take his PPK-bag containing a private Rolex GMT-Master, aboard LM "Challenger" down to the Taurus Littrow region, while he orbited the Moon aboard CSM "America".

This Rolex timepiece remained on the Moon for about 75 hours before humankind left the moon for good. Evans' personal Rolex Oyster Perpetual GMT-Master 1675 wristwatch was auctioned during the October 2009 Heritage auction in Dallas, Texas, and sold for $131,000 (US).

Meanwhile in the USSR

That same year (1971), Russian cosmonaut Vladimir Shatalov was pictured during Soyuz 10 training wearing a Rolex GMT-Master with red/blue Pepsi bezel. Four years later, when Shatalov trained as a backup for the ASTP mission (Apollo-Soyuz Test Project), he was still wearing that same GMT-Master. Moreover, Russia's third cosmonaut, Adrian Nikoyalev, was pictured wearing a Rolex GMT from 1963 to 1983.

Although it's not clear how these pilot watches ended up behind the iron curtain, some Soviet-Russian cosmonauts were pictured wearing a Rolex GMT-master wristwatch. Russia's third cosmonaut, Adrian Nikoyalev, was pictured wearing a Rolex GMT-master from 1963 to 1983.

1971 Apollo 14 Lunar Module Pilot Edgar Mitchell winding and calibrating his personal GMT-Master watches

ROLEX ON BOARD THE SHUTTLE AND ISS

With the almost routine planning of regular space shuttle flights, the rules for the types of personal items astronauts could carry aboard were relaxed.

Shuttle pilots recruited from military services always had an interest in watches. A good example being naval aviator Michael McCulley, who wore his personal Rolex Daytona 6263 aboard space shuttle Atlantis on STS-34 in October 1989.

Pilot astronauts, Sonny Carter and Blaine Hammond, respectively, wore a personal Rolex Datejust on STS-33 in November 1989, and a Rolex GMT-master Pepsi aboard STS-39 in May 1991. These were both classified Department of Defense missions aboard space shuttle Discovery.

Naval aviator Scott Altman wore four different Rolex watches in space. Rolex Submariner on Columbia STS-90 (April 1998), and GMT Masters on Atlantis STS-106 (September 2000), Columbia STS-109 (March 2002) and Atlantis STS-125 (May 2009).

In the latter half of the Space Shuttle era, one astronaut stood out by wearing his personal Rolex GMT-Master during each of his missions. NASA mission specialist, Dr. Leroy Chiao, wore his Rolex watch on three different space shuttle orbiters: Columbia (STS-65 in July 1994), Endeavour (STS-72 January 1996) and Discovery (STS-92 in October 2000).

In 2004, Chiao was also assigned as commander of the ISS Expedition-10 and was launched aboard Soyuz TMA-5 to the space station, wearing his trusted Rolex timepiece. During the 192-day mission, he became the first American citizen to vote in a presidential election while in space. Leroy Chiao accumulated a total of 229 days and 8 hours in space, bestowing upon his personal GMT-Master watch the claim of having the most flight time in space.

In April 1998 Canadian physician, Dafydd "Dave" Williams, wore a 1960s Rolex Submariner during the Columbia STS-90 mission. Being a keen scuba diver and NEEMO project aquanaut, he wore the same Submariner, a Rolex model with an epic diving history and pedigree, for several saturation dives on the Aquarius undersea research habitat.

In August 2007, Williams wore a Rolex Sea-Dweller during the STS-118 mission aboard the space shuttle Endeavour. The mission marked the 150th manned US space launch and Dr. Williams became

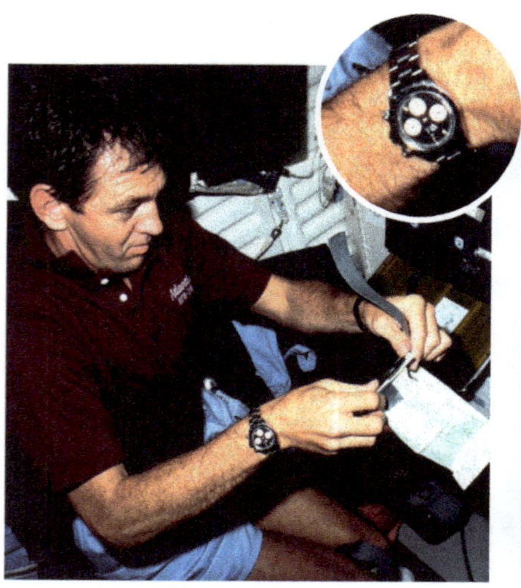

Michael McCulley with his Daytona 6263 aboard space shuttle Atlantis mission STS-34

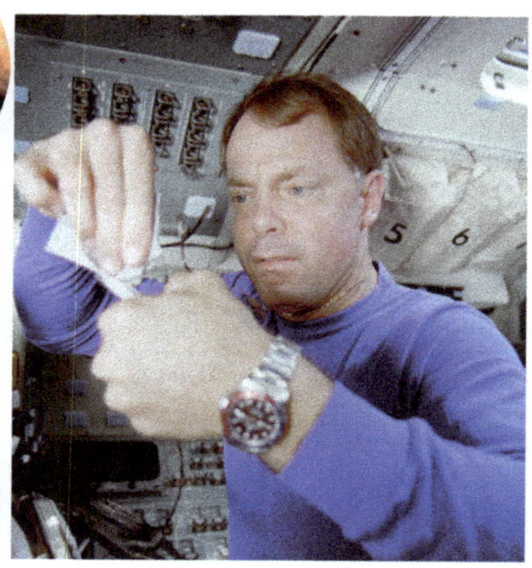

Lloyd Blaine Hammond pilot of the 1994 Space Shuttle Discovery mission STS-64

Dr Leroy Chiao and his GMT-Master on Space Shuttle Colombia STS-65 in 1994

the Canadian with the most spacewalks, passing Chris Hadfield in total EVA time.

In October 2007, Malaysian surgeon, Sheikh Muszaphar Shukor, took some of his personal watches, among them a Rolex Submariner, aboard Soyuz TMA-11 to the ISS. That same year Rolex had just released the certified chronometer version of the Submariner 14060M. This is still considered the last classic Submariner thanks to its clean dial, aluminum bezel insert, sapphire crystal, and case with drilled lugs.

Half a century after the release of Rolex aviation and dive watches, the GMT-Master, the Submariner, and the Sea-Dweller have earned the Professional Collection a rich heritage global reputation. This was earned at a time when mechanical watches were the pinnacle of advanced design and technology.

Rolex's commitment to the continuous improvement of their products took the form of a sustained focus on design for the harshest environments and the most rugged use by the most active customers. Rolex Professional watches have become synonymous with accuracy, durability, and amazing human achievement.

Sonny Carter with his Datejust

1998, Scott Altman onboard space shuttle Columbia STS-90 with his Submariner

Scott Altman on the Flight Deck of Shuttle Atlantis mission STS-125 in 2009 with his Two-Tone GMT-Master "Rootbeer" Nipple Dial

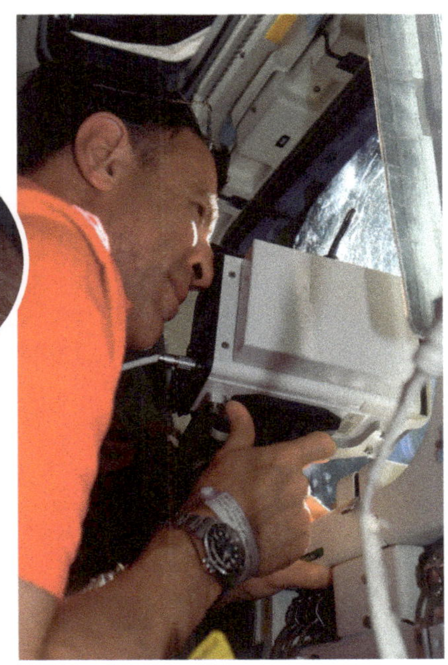

2007, Dafydd "Dave" Williams onboard space shuttle Endeavour STS-118 with his Sea-Dweller

OYSTER PERPETUAL SUBMARINER

Launched in 1953 and presented at the Basel Fair the following year, the Submariner was the first wristwatch in the world to be waterproof to 100m (330ft). It was not, however, the first watch designed for diving. This honor goes to the Panerai and then Blancpain Fifty Fathoms with the first rotating bezel.

The Submariner's rotating bezel is designed for measuring elapsed time underwater. While not an industry first, it is elegantly executed by Rolex. It was this simple feature, and not the water resistance, which it shared with other collections, that marked the Submariner as a professional tool watch.

Introduction of the date feature in 1969 made the Submariner a mainstream watch, appealing to a broader and more aspirational market. Purists still consider the no-date Submariners to be the most faithful to the original design concept, and shun the date window and magnifying cyclops lens.

The first Submariner references are the 6200, 6204 and 6205, which were small by today's standards (38mm). These earliest examples are some of the hardest to validate and authenticate due to decades of wear, numerous owners, and a large number of minor variations in their configurations.

Like vintage cars of the same era, most of these early references have had extensive work to keep them running. This makes the widespread obsession with originality and authenticity subjective and dogmatic.

There are countless tiny variations in the four-digit reference series of Submariners. Some had depth ratings printed meters first units for European markets, while others had feet-first units of measurement.

Some models had depth ratings printed in red or Explorer-style 3-6-9 hour markers. While the 3-6-9 style was intended for military orders, they made it to the commercial market and proved popular. Other dial variations include gilt text, nipple indices, open and closed minute tracks, and gloss and matte lacquer finishes. Some had pencil-style hands, others had Plongeur style (military orders only). Early bezels were friction fitted and bi-directional while later versions became clicking and unidirectional. It's this sheer variety that makes Submariners so collectible.

Bezels were bi-directional until 1953, thanks to a patent owned by Blancpain for their Fifty Fathoms watch. Once the patent expired, the Submariner gained a unidirectional bezel, beginning with the 55xx series.

High demand and rising prices have encouraged unsavory practices, making it challenging to find an honest example. Buying poor examples for restoration is a costly, risky, and lengthy proposition. These ill-advised projects of passion seldom make economic sense. If you're motivated to take one on, expect restoration costs to match the already astronomical acquisition cost – unless you're equipped to do the work yourself.

Modern Classic iterations of the no date Submariner, such as the 14060 continue to be popular for daily wear. The clean dial symmetry and contemporary features make these watches attractive, practical, and collectible. Prices are softer than their four-digit predecessors like the ref. 5513 and 5512.

The only four-digit reference with a date complication is the ref. 1680. This is a complex and long-running reference with seven dial variations during its lifetime. Only the ref. 1680 and 16800 (matte dial variation) qualify as vintage. The use of the vintage label is controversial for any reference that followed the matte 16800.

Bezel inserts are an essential part of the value and price of any Steel Sport Watch. Service replacement bezels are identified by their thin font printing. Bezels with a large and broad print font are considered original and correct for the ref. 1680.

Faded (ghost) bezels should be evaluated against the condition of the case, dial, and hands; i.e., an aged bezel should be paired with an

SUBMARINER MOVEMENTS

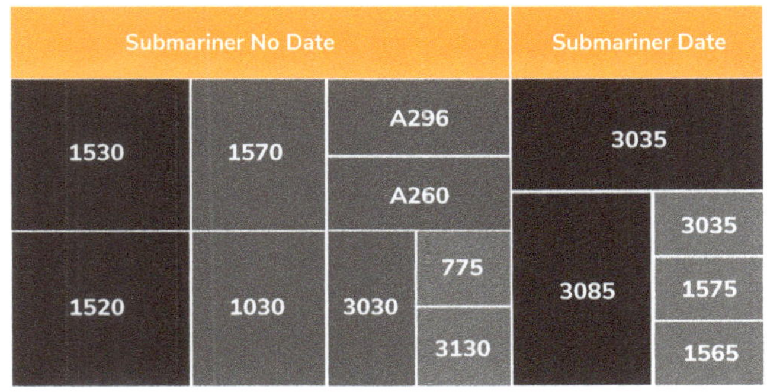

equally old case, dial, and hands. Buyers should be looking for even aging on the whole watch.

Given the price premium of a well-worn and faded bezel insert, unscrupulous dealers are known to artificially age bezels with bleach and high temperature (a practice known as "cooking a bezel"). You should avoid watches with mismatched wear patterns.

Of the 30 Submariner references, 13 have a date function and 17 have no date. Twelve different movements (or calibers) have been used across the line since 1952, with the caliber 3135 becoming the most common.

With so many references and variations within references, I will make individual notes on only those examples I believe to be most interesting to most new collectors.

Submariner dial ref. 6204 with meters-first gilt dial and closed minute track

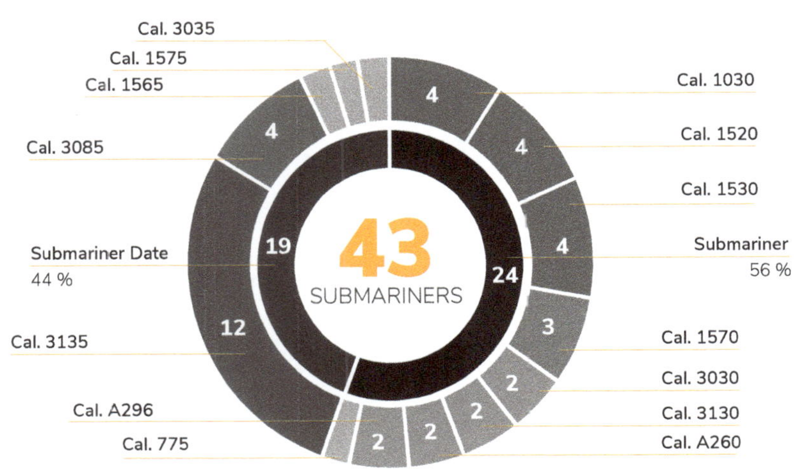

Submariner ref. 6538

SUBMARINER EVOLUTION TIMELINE

1954
ref. 6204 Debut of the Submariner at the Basel Fair

1955
ref. 6538 Successor to 6204 Pre-Sub, with caliber 1030 movement.
ref. 6536 Successor of reference 6205 with big crown.
ref. 6536/1 Chronometer version of reference 6536 with caliber 1030.
ref. 6538 British Royal Navy choses the Submariner.

1956
A new design for the hands and the bezel gets 15 minute index markers.
ref. 6538/8 The 6538 receives the same case as reference 6200.
ref. 6536 A red triangle on the bezel to mark the zero position.

1958
Font change on the bezel with 0 (zeroes) becoming square-ish
ref. 5510 Based on reference 6200 with caliber 1530 (1957)
ref. 5508 6536/1 with caliber 1530.

1959
ref. 5512 Crown guards, increased case diameter to 40mm from 36mm, "Superlative Chronometer, Officially Certified"
ref. 6538 "Superlative Chronometer, Officially Certified".

1962
ref. 5513 Updated crown guards and upgraded cal. 1530.

1963
ref. 5513 Updated with cal. 1520 introduced in the same year.

1966
ref. 1680 Date window, plexi crystal with cyclops, red writing on the dial (till 1973) and cal. 1575.

1969
ref. 16618 Submariner available in 18k gold.

1979
ref. 16800 Submariner receives a sapphire crystal and water resistance is increased to 300 meters.

1981
ref. 168000, 16610
The Submariner receives a unidirectional bezel so remaining diving time can only be shortened. Caliber 3085 is adopted..

1983
ref. 16613 The Submariner becomes available in "Rolesor", a two tone gold and steel.

1988
ref. 16610 Upgraded with caliber 3135.

1990
ref. 14060 with caliber 3000.

2002
ref. 14060M, last of the classic two-line Submariners updated with caliber 3130.

2003
ref. 16610 LV
A green bezel is fitted to a 50th Anniversary Submariner which came to be known as 'the Kermit'..

2009
ref. 16613 LB
First Submariner fitted with a ceramic bezel, in rolesor (gold and stainless steel). Case lugs became fat and new bracelet added..

2010
ref. 16619
White gold Submariner with blue dial and ceramic bezel.
ref. 116610LN/LV
The LV is introduced with a green dial and green ceramic bezel..

1960 1970 1980 1990 2000 2010

Modern Classic Submariner Date 16800, with COSC cal 3035

Modern Classic Submariner 14060

Modern Classic Submariner Date 16610

Vintage 1956 Submariner 6536

SUBMARINER REF. 5510

SUBMARINER REF. 5512

The Submariner 5512 appeared in 1959 and was made for about 15 years. During that time it went through several variations, most notably, both a COSC and non-COSC version (four versus two line specification text). The four-line COSC version has the text 'Superlative Chronometer Officially Certified' printed in a distinctive second color. Other dial variations include gilt and white markers with open and closed chapter rings (a circle enclosing the minute markers).

This was the first reference to have crown guards. The earliest versions were the pointed-style crown guards. These proved to be marginally effective, and were subsequently upgraded and beefed-up to the more familiar rounded style. This case innovation was developed and perfected in the 5512.

These examples illustrate some of the variations including the rare Exclamation Point (!) dial, denoted by the round lume dot beneath the 6 o'clock hour marker.

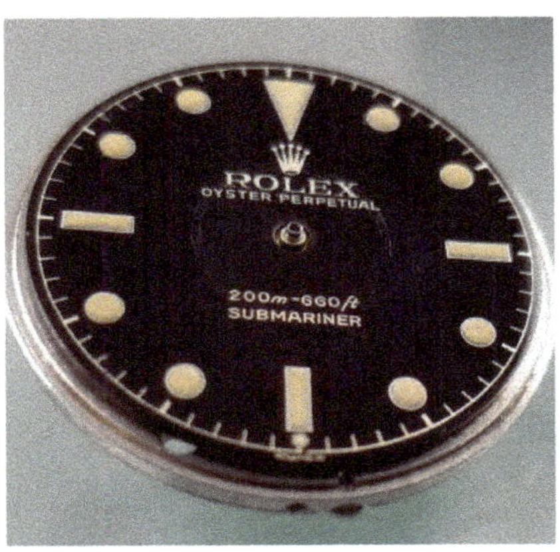

SUBMARINER REF. 5513 (NO DATE)

The ref. 5513 was made from 1962 until 1990, one of the longest production runs of any Rolex. As a result, there are many good examples in circulation. Like other long-running references, there were many minor variations as the model evolved. These included two different movements, the cal. 1530 and cal. 1520.

This reference has a strong James Bond 007 connection, which appeals to some collectors. George Lazenby wore a ref. 5513 with an Oyster bracelet in parts of On Her Majesty's Secret Service, as did Roger Moore in Live and Let Die and The Man with the Golden Gun.

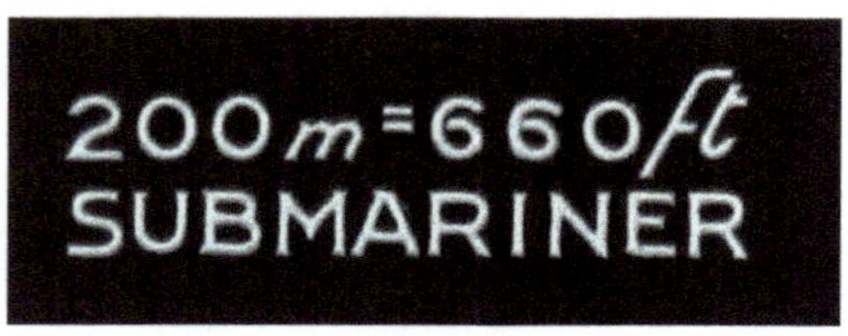

The ref. 5513 appears with two lines of text below the center point, indicating this reference was not COSC certified, i.e. not chronometer-grade as determined by Controle Officiel Suisse des Chronomotres, or the Official Swiss Chronometer Testing Institute.

The ref. 5512, on the other hand, had four lines of text and was COSC certified. The 5512 features the Superlative Chronometer, Officially Certified text below the Submariner and depth rating. Production of the two references (5512 & 5513) overlapped.

The 5513 had three genres of dial style, with several variants. The earliest iterations had gilt gloss dials from 1962–1966. The desirable matte dials and maxi variants were made from 1966 to around 1984, before reverting back to a gloss dial (with new white gold indexes).

Within the matte maxi dial genre, five distinct variants have come to be known as MK I to V. The distinctions between them come down to minor differences in text font and coronet shape, and can require a loupe to detect.

Meters-First matte dials started in 1966 around the 1.6M serial range and ran to around 1968. The tritium lume application on these is known to be inconsistent and variable. The early versions had thin layers of lume that can be quite hard to detect and may require a loupe. In the middle of the production run, the lume appears to be applied more thickly and has a domed appearance typical of hand-applied lume markers. By the end of this dial series, the tritium has a flat, more even appearance that's more consistent with pad printing techniques.

Non-serif dials followed in 1969 and ran to 1970, starting in the 2M serial range. Serif dials followed these, running until 1973 with serials in the 3M range. After that, things returned to the non-serif style until around 1976.

Maxi dials started after 1976 and ran to 1984, and were subsequently replaced by the gloss dial with the white gold hour markers.

You can use this timeline to check the dial is an appropriate style and matches the expected serial number of a watch.

SUBMARINER REF. 5513 DIAL GENRE TIMELINE

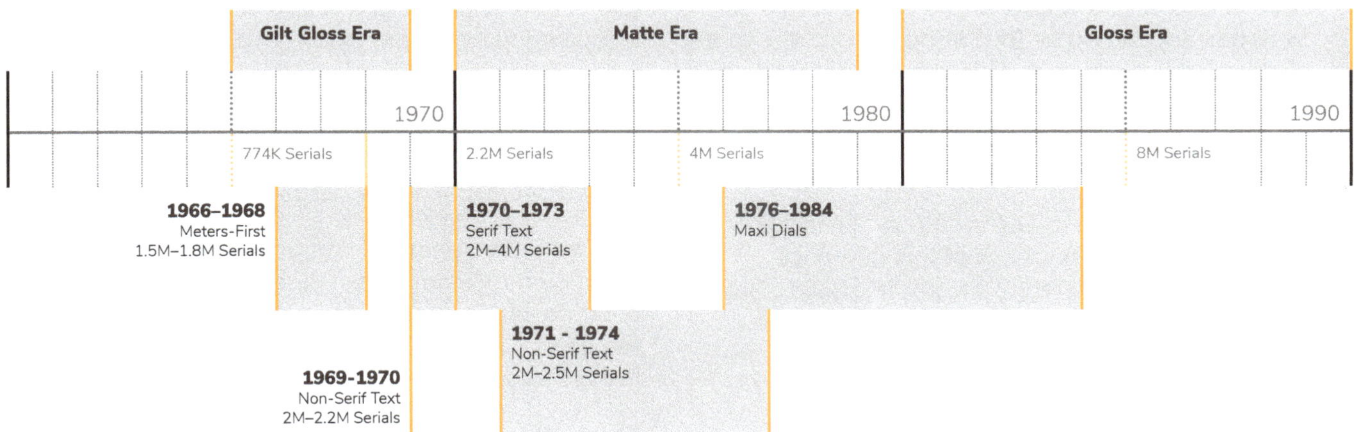

Extract from the Rolex Submariner complimentary booklet with each new watch

RED SUBMARINER DATE REF. 1680

The 'Red Submariner' is one of the most desirable and collectible variations of the ref. 1680. These watches are known as Red Submariners due to the distinctive red dial text. This is a complicated and expensive reference with many potential pitfalls. If you are new to vintage Rolex, it is inadvisable to take this reference on as your first acquisition.

This hallmark red text was eventually replaced with white in later iterations of this reference. These are often referred to as 'White 1680s'. The presence of a single line of red text commands a substantial price premium and an almost obsessive and rabid following.

Collectors continue to debate whether the ref. 1680 is the most desirable vintage Submariner of all time or just the most desirable featuring a date complication. Personal preference plays a large part in this debate.

It is common for white and red text dials to be swapped, meaning that a red text dial may be found in a later case that is out of range of the expected serial numbers for Red Subs, or a white text dial may be found in a case which formerly belonged to a Red Sub. Given the value of an authentic Red Sub, collectors should assess the watch in microscopic detail and be aware of promiscuous dials.

Until a Red Sub can be examined in hand by a prospective buyer or an acknowledged Red Sub expert, a potential buyer should assume the watch is not authentic. In this case, obtain a commitment of 100% authenticity from the seller and written details of their return policy. Buyers will need time to have the watch appraised and some recourse if undisclosed issues are identified.

Of the Red Submariners, there are seven dial variants including a Luminova service dial. All seven appear concurrently and are overlapping during three years from the introduction in 1969 to 1972. The variations are minute and require good macro photos or a loupe to examine.

There is also a Red 1680 service dial that is identifiable with the "SWISS" marking at 6 o'clock, indicating use of Luminova. These still glow.

Some dials of early 1680s (Mark I, II, & III) are prone to turning brown, giving rise to the 'Chocolate Submariner' moniker. Brown dials are a result of paint defects and highly desirable and collectible.

The red Submariner printing on Mark I, III, and IV are known to be imperfect. It is not uncommon for white to appear around the lettering or through the red paint. These imperfections are recognized by the collector community and accepted as legitimate.

Casebacks from the range 2M to about 3.4M were stamped with a date code starting with II 69 (second quarter of 1969). This practice appears to end with II 72 in the 3M range. Watches from 1973 onward ,and Rolex service replacements, do not have a stamped date code on their casebacks.

The movement used in the ref. 1680 is the caliber 1575 (the trailing 5 denoting the date complication). It is common for the auto winding rotor to be stamped 1570, as this part was interchangeable between the Submariner Date and the no-date Submariner.

This movement was also used in other less desirable and inexpensive references. Verifying that the movement is a cal.1575 doesn't mean it was the original one shipped from the factory. Only Rolex can match a case serial to a movement serial number and confirm they left the factory together.

The ref. 1680 has variation in the warranty papers it was sold with. The most desirable and valuable warranty papers are punched. Handwritten or blank documents carry little value or price premium.

Submariner Red MK I dial text

MK1 dial with meters-first depth rating. These appear in the serial range, approximately 2.07M to 2.2M or roughly 1969 to 1970. They use a thick, heavy font with closed 6's. Note the distinctive long curve of the letter 'f'.

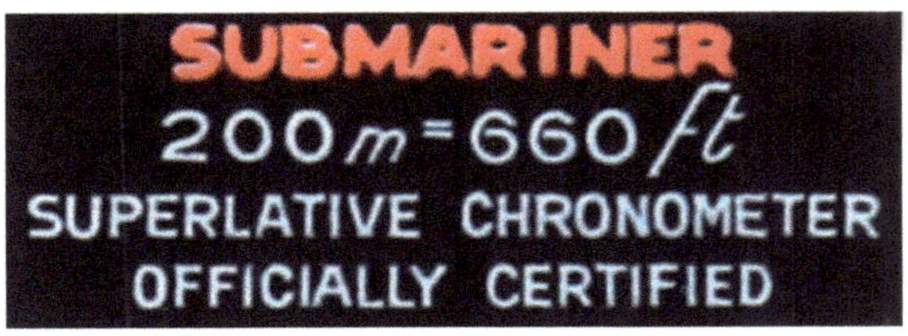

Submariner Red MK II dial text

MK II dial with meters-first depth rating. Serial range approximately 2.2M to 2.45M or 1970. It has a lighter, thinner font with open 6's. Note the sharper angular curve of the letter "f".

Submariner Red MK III dial text

MK III dial with meters-first depth rating. Serial range also in the 2.2M to 2.45M range (1970) and overlaps with the MKII. Uses open 6s but with a thicker font. Red lettering is over-printed on the white text below. Note the shorter top of the letter "f".

Submariner Red MK IV dial text

MK IV dial with feet-first depth rating. Serial range in the 2.45M to 3M range or 1970 to 1972. Red lettering is over-printed on the white text below. Note the distinctive open 6s of the depth rating.

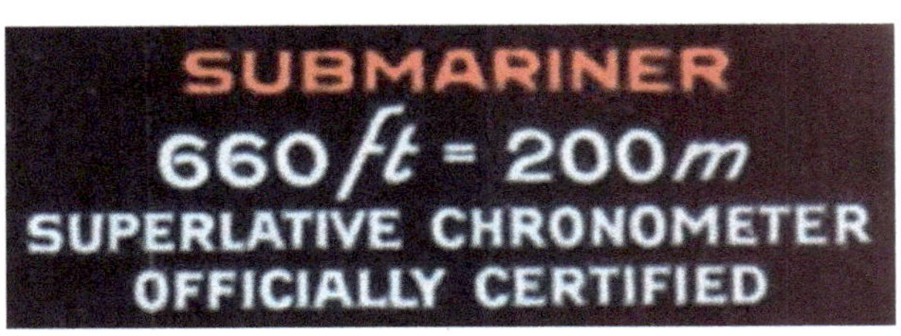

Submariner Red MK V dial text

MK V dial with less pronounced but still open 6s. Serial range 2M to 3M or 1970 to 1972. Red text is directly printed on the dial and not over-printed on the white text below. Note the flatter, almost box-shaped printing of the letter 'S'.

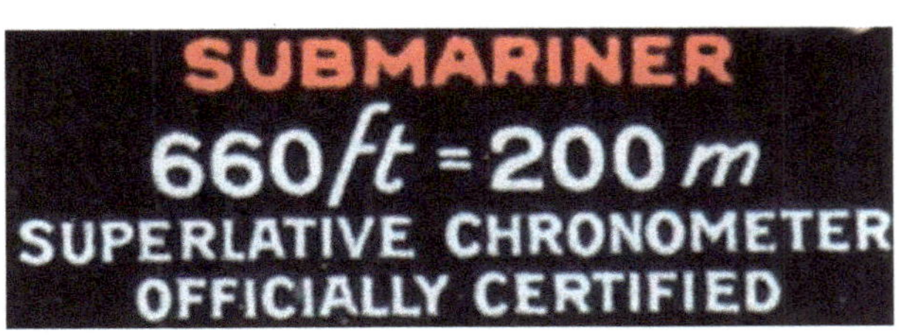

Submariner Red MK VI dial text

MK VI dial with closed 6's and a rounded fuller font with the 'S' more rounded. Seen in the 3M to the 4M serial range and gone by 1975.

ROLEX
SUBMARINER
1000ft=300m

WHITE SUBMARINER DATE REF. 1680

The 'White Submariner' has not achieved the cult status or the prices of the red text predecessor. There are three acknowledged variations of the white dial ref. 1680. All three are found in watches across the serial range.

The MK I dials are made by Lemrich.

The back of the dial is stamped with a number with a 121 prefix. Note the 'L' in "ROLEX" is printed directly below the coronet, and the '=' in the depth rating is below the 'A' of "SUBMARINER". Also note the alignment of the bottom row of "OFFICIALLY CERTIFIED" text and how it differs from the MK III.

The MK II dials are made by Beyeler.

The back of these dials had repeating text, "Beyeler Geneve" stamped across them. Note that the letter 'L' of "ROLEX" is shifted left below the coronet.

The MK III dials are also made by Lemrich, like the MK I.

The shifting of the 'L' of ROLEX is more pronounced and the position of the '=' relative to the space between "OFFICIALLY" and "CERTIFIED" appears different from the MK I.

A typical MK II dial

THE SOLID GOLD SUBMARINER REF. 16808

The solid 18K yellow gold Submariner may be an acquired taste but it is undoubtedly a statement piece. The yellow gold is a heavy dominant presence on the wrist and said to scratch and scuff easily, and the gold Oyster bracelet is prone to stretching. Launched in 1977, the gilt nipple dial version is the most desirable and collectible.

GREEN SUBMARINERS (HULK & KERMIT)

Rolex is extremely conservative in its use of color, particularly its signature green. Green is reserved for anniversary celebration models. The 50th anniversary of the Submariner was celebrated with a special edition of the ref. 16610. If you're looking for something a little different, these green Submariners, affectionately known as 'Hulk' and 'Kermit', are excellent choices.

REF. 16610LV KERMIT (2003 TO 2010)

Developed to celebrate the Submariner's 50th anniversary, the Kermit, as it has come to be nicknamed, is distinguished by the green bezel insert. It is also the first instance of a Submariner with a glossy maxi dial (enlarged hour indices). This version has slightly wider hands than the standard black-dial ref. 16610, though you'll need to be experienced with both to notice the difference.

The first models were issued in the autumn of 2003 on a "Y" serial in September, with low "F"s following in October through December. Because of this, only a narrow range of watches are true anniversary models issued on the Submariner's 50th birthday.

The Oyster case of the Kermit has no drilled lug holes. Some collectors claim to have seen early press pictures of the Kermit with drilled lug holes, and anecdotal evidence suggests prototypes had drilled lugs – as did the first production batch given to Rolex executives. Rumor has it these photos were released to throw off counterfeiters.

The Kermit was upgraded in 2010 with the six-digit 116610LV. Now out of production, it has become sought after and collectible as a future classic.

The Kermit commands a price premium over the standard black bezel version (16610) of the same year. If pursuing one of these, pay careful attention to the authenticity of the green bezel and the size (width) of the hands. It is possible to swap a black bezel on a standard 16610 for a green one (possibly aftermarket) and pass it off as a 16610LV Anniversary Model with a price premium. A bezel swap is a simple undertaking requiring little more than a knife and masking tape. Swapping hands is possible too, but is a more involved job requiring specialized watchmaker's tools.

The "LV" designation should be indicated on the warranty card, along with the case serial number. The earliest versions, with the old style punched warranty papers, were apparently only marked with a "V" designation.

If you're going to buy a 16610LV, you are strongly advised to have it verified by an RSC or AD with a Rolex-trained watchmaker on staff.

REF. 116610LV HULK (2010 TO PRESENT)

The Hulk is an evolution of the anniversary Kermit, and the sunburst green maxi dial is a significant change from the five-digit Kermit it replaced. It has the new Cerachrom bezel and the broader lugs of what has become known as the modern maxi case. It wears larger on the wrist and has an aggressive masculine character.

REF. 116618LB BLUESEY

The modern solid 18k yellow gold Submariner is the final execution of the original tool watch. As if solid gold was not ostentatious enough, the polished center links on the Oyster bracelet provide the extra visual pop, making this reference immediately recognizable as a truly luxury watch. The sunburst blue dial responds to natural sun light in a distinct and unique manner, appearing in shades from deep midnight blue to pale sky blue.

These pieces are popular among collectors with big, self-confident personalities, and muscular forearms and wrists to match.

MILITARY SUBMARINERS

The subgenre of Submariner with military provenance is complex. Rolex watches purchased by the military are state property and issued to military personnel along with other specialized tools such as diving equipment. Like all military equipment, these government-owned watches bear military engravings and serial numbers. They never became the personal property of the servicemen who used them and any failure to return them to the quartermaster after use could have severe consequences. However, some watches that survived active duty eventually made their way into private ownership via military surplus auctions.

Government-issued examples are highly prized and should not be confused with those purchased privately by servicemen during their active service. These personal purchases usually took place at exclusive retail outlets (often tax-free) such as the British NAAFI (Navy Army Air Force Institutes) and the US China Fleet Club. While their owners bestow a military provenance on these watches, they are not Military Submariners in the strict sense.

The third class of military Submariner are those watches gifted to soldiers by a foreign sovereign. Examples include the red Khanjar or Qaboos Sea-Dwellers, given by the Sultan of Oman to 90 Special Air Service (SAS) soldiers after the Battle of Mirbat in 1972. These unique gifts of appreciation have a military provenance unlike the government-issued examples. Examples like this are pursued by both watch nerds and military history buffs.

The Ministry of Defense (MOD), is the official customer and first owner of British Military Submariners

Rolex supplied special issue Submariners to the British Ministry of Defence. These watches were for use by special forces of the Royal Navy and British Army, specifically the Naval Special Boat Squadron (SBS) and the Army Special Air Service (SAS). Tudor Submariners were issued to other elite military forces including the French, Canadian, and Australian armed services as well.

The term 'Milsub' usually refers to only four specific Rolex references – ref. A/6538 (or 6538A), 5513, 5514, and 5517. The most collectible and desirable is the ref. 5517, produced in limited volume. The ref. 5517 is a rare and valuable piece, which means they are also frequently counterfeited to a very convincing quality. Thankfully, Milsubs are relatively well cataloged and documented, and Milsub specialists can verify them. Given the nature of the work performed by their owners, these records are closely held.

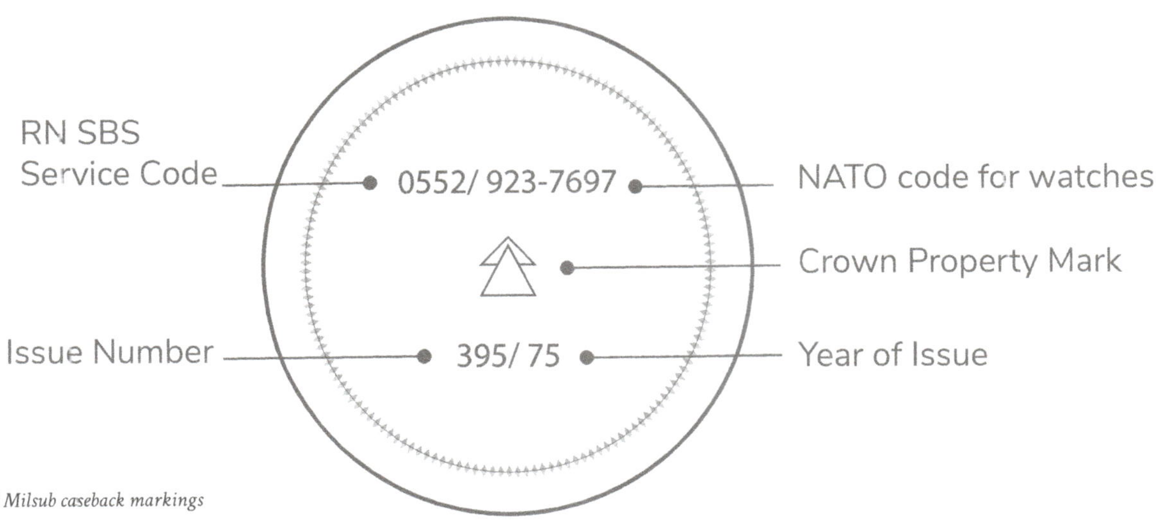

Typical Milsub caseback markings

Ian Flemming

007 AND THE 6538

Ian Lancaster Fleming (May 28, 1908 – August 12, 1964) served in the British military during World War II in Naval Intelligence. He wrote 14 James Bond novels, nine 007 short stories, and Chitty Chitty Bang Bang! He wrote them all from his Caribbean island home in Jamaica, called Golden Eye.

Ian Fleming wrote that his James Bond character wore a Rolex like he himself, who wore a stainless steel Rolex Explorer 1016 and a Datejust. He spent much time consulting on the set of the first three James Bond films, including Dr. No (1962), From Russia With Love (1963), and Goldfinger (1964).

Sean Connery starred in these first three 007 movies with the iconic Rolex Submariner 6538. His watch had a striped regimental NATO strap, which was distinctive and unusual for the time.

The Royal Scots Regiment, also known as the Royal Regiment, is the oldest British Army Regiment. Their regimental colors are Peony Red, Gosling Green, and Oxford Blue.

In subsequent years, 007 was seen wearing a Submariner 5513 and an early Cosmograph (George Lazenby). This was followed in the 1970s and 80s with a Submariner Date 1680 and a Datejust (Roger Moore).

REF. A/6538 (1957 TO 1966)

The first orders for special issue Submariner came from the British Royal Navy from 1957 to 1966. These were for a modified version of the James Bond Goldfinger ref. 6538.

The ref. A/6538 had modifications including a larger (and taller) bezel made out of nickel silver instead of the 6538's plated brass. This modification allowed frogmen better grip with wet, gloved hands. It also had fixed, 2mm spring bars. Engravings on the casebacks include the war department arrow, military branch, and year of issue. From 1967 to 1971, the British Navy switched suppliers and adopted the Omega Seamaster 300.

Rolex sold a stock Submariner ref. 6538 to the Royal Canadian Navy beginning in 1956/7. These versions have an identification and service number engraved on the inside of the caseback. The configuration is otherwise the same as the standard production Rolex Submariner.

007 regimental NATO strap on the Submariner 6538.
Sharp eyed observers will note that is 4mm too small for the watch.

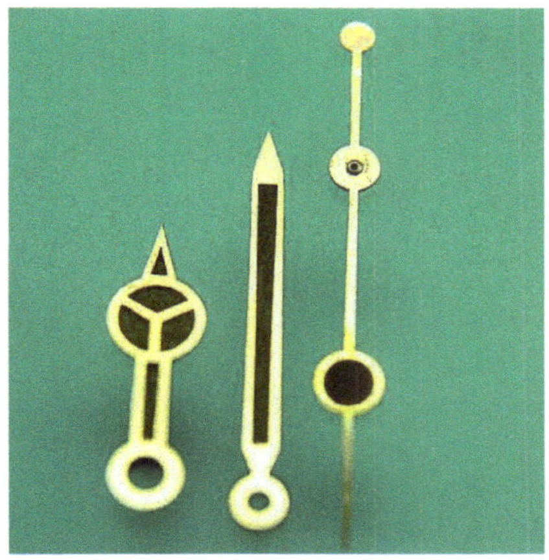

200m = 660ft
SUBMARINER

PATENTED

MONTRES ROLEX SA
GENEVA
SWITZERLAND
PATENTED
STAINLESS STEEL
6538

REF. 5513 & 5517 (1972–1979)

The second period of Milsubs was for modified refs. 5513 and 5517, and Rolex supplied approximately 1200 units. About 180 of these examples are accounted for and attributed to known collectors.

The first batch of ref. 5513 had case engravings between the 12 o'clock lugs. The second batch bore the same 5513 case engravings but with an additional 5517 stamp on the underside of the seven o'clock lug. The third set bore the 5517 case engravings between the lugs. This third batch has the more legible, broadsword, or plongeur-style hands and an arrow style second hand. Rolex did not make these hands, and during servicing, they were often replaced by the Rolex Mercedes-style hands (Squelette).

All three batches featured a matte dial with a 600ft = 200m depth rating and a circled "T", denoting the use of Tritium lume. All of their bezels had continually graduated markers for the full 60 minutes and an anti-reflective satin case finish. The case had fixed bars to enhanced durability. These fixed bars allowed for use with Nato straps and could be affixed to a "dive board."

All the ref. 5513 and ref. 5517 casebacks have the military, Crown Property mark, and unique issue numbers. Casebacks for the Royal Navy SBS versions have a UK Ministry Of Defence (MOD) service code of 0552 along with a NATO code for dive watches of 923-7697. They will also have a Broad Arrow-style insignia with an issue number and year of issue.

Casebacks for Army SAS versions have a MOD service code of W10. They also have three NATO codes – 6645, denoting a timing instrument; 99 for the UK; and 923-7697 for divers' watches. They will also have an issue number and a year of issue.

REF. COMEX 5514

The ref. 5514 was made under contract specifically for Comex, a French marine engineering company specializing in deep water saturation diving, and was never sold directly to armed forces of any foreign governments. The ref. 5514 was a modified ref. 5513 with a helium escape valve (HEV) to allow safe decompression of watches after saturation dives. Without an HEV, crystals were known to burst off the case during decompression.

Its designation as a Milsub is controversial and based on one example that came up for sale at Christie's Auction House. It had a dial marked "ARA", which stands for Armada de la Republica Argentina, or Argentine Navy. It had an ARA military issue number 68507-Ci.

Researchers at Christie's confirmed that Comex commissioned this 5514 along with other Comex dialed ref. 5514 watches. Comex requested the inscription "ARA" above the depth rating on the dial instead of their "COMEX" logo.

There were ARA-dialed ref. 5514s issued to 16 Argentine naval divers. These servicemen were training at the Comex Hyperbaric Experimental Centre in Marseille, France in 1977. It is likely that the Argentine military engraved issue numbers on the casebacks upon their return to duty in Argentina.

SUBMARINER REFERENCES

Name	Ref	Cal	Depth Rating	Start	End	Bezel	Size	Case	Description
Submariner No-Date	**6200**	775	200m/660ft	1953	1954	Numbered, LN	36mm	SS	Big 8mm crown with Explorer-style (3-6-9) hour markers on a gilt dial. May not feature "Submariner" branding 200m depth rating with fatter case. Considered the very first Submariner
Submariner No-Date	**6204**	A260	100m/330ft	1953	1954	Numbered, LN	36mm	SS	100m depth rating with thinner case than 6200. Pencil style hour hand. Seen with black honeycomb gilt dial
Submariner No-Date	**6205**	A260	100m/330ft	1953	1957	Numbered, LN	36mm	SS	Black honeycomb dial 100m depth rating with thinner case than 6200 Early version had pencil hour hand and later versions the mercedes hour hand Also seen without crown guards
Submariner No-Date	**6200**	A296	200m/660ft	1953	1955	Numbered, LN	36mm	SS	Big 8mm crown with Explorer-style (3-6-9) hour markers and gilt dial. May not feature "Submariner". Fatter case for improved 200m depth rating. Considered the very first Submariner
Submariner No-Date	**6202**	A296	100m/330ft	1953	1954	Numbered, LN	36mm	SS	Pre-Submariner Occasionally branded "Monometer"
Submariner No-Date	**6536**	1030	100m/330ft	1954	1958	Numbered, LN	36mm	SS	No crown guards. May feature Explorer 3-6-9 hour markers with two or four line colored print ,such as red depth markings. Waterproof to 100m
Submariner No-Date	**6538**	1030	200m/660ft	1957	1960	Numbered, LN	40mm	SS	First to be rated to 66oft/200m. First to have Triplock Big Crown (8mm). Bezel introduced 15 minute markers and red triangle at 12 o'clock. The James Bond Dr No ref. First use of Mercedes hands
Submariner No-Date	**5508**	1530	100m/330ft	1958	1965	Numbered, LN	36mm	SS	May feature Explorer (3-6-9) hour markers with two or four line colored print such as red depth markings. May feature big or small crowns. Both COSC and non-chronometer movements. Gilt dial
Submariner No-Date	**5510**	1530	200m/660ft	1958	1960	Numbered, LN	36mm	SS	Transitional reference, may feature Explorer 3-6-9 hour markers and depth rating to 200m. Two or four line colored print such as red depth markings. Fixed spring bars indicate military issue Big crown (8mm)
Submariner No-Date	**5510**	3030	200m/660ft	1958	1960	Numbered, LN	36mm	SS	Transitional reference, may feature Explorer 3-6-9 hour markers and depth rating to 200m Two or four line colored print such as red depth markings Fixed spring bars indicate military issue Big crown (8mm)
Submariner No-Date	**5512**	1520	200m/660ft	1960	1980	Numbered, LN	40mm	SS	First Sub with crown guards. Later versions were COSC "Superlative Chronometer, Officially Certified"
Submariner No-Date	**5513**	1520	200m/660ft	1960	1990	Numbered, LN	40mm	SS	Some Royal Navy and some COMEX. Dials change from gilt gloss to meters-first matte dials around the 16M serial range or 1966
Submariner No-Date	**5514**	1520	200m/660ft	1960	1978	Numbered, LN	40mm	SS	Non-COSC COMEX
Submariner No-Date	**5517**	1520	660ft/200m	1960	1978	Numbered, LN	40mm	SS	Military Issue Only ordinance markings. Plongeur sword hands
Submariner No-Date	**5512**	1530	200m/660ft	1960	1980	Numbered, LN	40mm	SS	First Sub with crown guards. Later versions were COSC "Superlative Chronometer, Officially Certified"
Submariner No-Date	**5513**	1530	200m/660ft	1960	1990	Numbered, LN	40mm	SS	Some Royal Navy and some COMEX. Dials change from gilt gloss to meters first matte dials around the 16M serial range or 1966
Submariner No-Date	**5512**	1570	200m/660ft	1960	1980	Numbered, LN	40mm	SS	First Sub with crown guards Later versions were COSC "Superlative Chronometer, Officially Certified"
Submariner No-Date	**5513**	1570	200m/660ft	1960	1990	Numbered, LN	40mm	SS	Some Royal Navy and some COMEX. Dials change from gilt gloss to meters first matte dials around the 16M serial range or 1966

SS : Stainless Steel | YG : Yellow Gold | WG : White Gold | LN : Lunette Noir

Name	Ref	Cal	Depth Rating	Start	End	Bezel	Size	Case	Description
Submariner No-Date	**5514**	1570	200m/660ft	1960	1978	Numbered, LN	40mm	SS	Non-COSC COMEX
Submariner No-Date	**6540**	1030	660ft/200m	1971	1978	Numbered, LN	40mm	SS	UK Ministry of Defence issue Plongeur sword hands, 60 min bezel markers (approx 1200 units made including 6538A)
Submariner No-Date	**6538A**	1030	200m/660ft	1971	1978	Numbered, LN	40mm	SS	UK Ministry of Defence issue Plongeur sword hands, 60 min bezel markers (approx 1200 units made including 6540)
Submariner No-Date	**14060**	3030	1000ft/300m	1989	2012	Numbered, LN	40mm	SS	Introduction of sapphire crystals, white gold surrounds on hour markers. Tritium lume ageing nicely
Submariner No-Date	**14060M**	3130	1000ft/300m	2000	2012	Numbered, LN	40mm	SS	A transitional reference with uprated movement as denoted by "M". Last reference to feature drilled lugs
Submariner No-Date	**114060**	3130	1000ft/300m	2012		Numbered, LN	40mm	SS	Introduction of Super Luminova and the engraved rehaut
Submariner Date	**1680**	1565	660ft/200m	1969	1981	Numbered, LN	40mm	SS	Chronometer Red & White
Submariner Date	**1680**	1575	660ft/200m	1969	1981	Numbered, LN	40mm	SS	Chronometer Red & White
Submariner Date	**1680**	3035	660ft/200m	1969	1981	Numbered, LN	40mm	SS	Chronometer White only
Submariner Date	**16800**	3085	660ft/200m	1977	1987	Numbered, LN	40mm	SS	Upgraded 1680
Submariner Date	**16803**	3085	660ft/200m	1977	1987	Numbered, LN	40mm	SSYG	Two tone 16800
Submariner Date	**16808**	3085	660ft/200m	1977	1987	Numbered, LN	40mm	YG	Gold 16800
Submariner Date	**16618**	3085	1000ft/300m	1987	2010	Numbered, LN	40mm	YG	18k gold version of the 16610
Submariner Date	**16610**	3135	1000ft/300m	1987	2010	Numbered, LN	40mm	SS	Successor to the 16800 and 168000. Final Sub with aluminum bezel and first Sub to adopt Luminova and etched rehaut
Submariner Date	**16613**	3135	1000ft/300m	1987	2010	Numbered, LN	40mm	SSYG	Two tone version of the 16610
Submariner Date	**16618**	3135	1000ft/300m	1987	2010	Numbered, LN	40mm	YG	18K gold version of the 16610
Submariner Date	**168000**	3135	1000ft/300m	1987	1988	Numbered, LN	40mm	SS	Transitional reference, first to feature 906L stainless steel
Submariner Date	**16610LV**	3135	1000ft/300m	2003	2010	Numbered, LV	40mm	SS	Kermit with green bezel and black dial
Submariner Date	**116610LN**	3135	1000ft/300m	2010		Numbered Cerachrom, LN	40mm	SS	First cerachrom bezel and maxi dial and case. Slightly wider minute hand and Chromalight lume
Submariner Date	**116610LV**	3135	1000ft/300m	2010		Numbered Cerachrom, LV	40mm	SS	Hulk with ceramic green bezel and dial
Submariner Date	**116613LB**	3135	1000ft/300m	2010		Numbered Cerachrom, LB	40mm	SSYG	Two tone with blue bezel and dial
Submariner Date	**116613LN**	3135	1000ft/300m	2010		Numbered Cerachrom, LN	40mm	SSYG	Two tone 116610
Submariner Date	**116618LB**	3135	1000ft/300m	2010		Numbered Cerachrom, LB	40mm	YG	Gold 116610 blue dial
Submariner Date	**116618LN**	3135	1000ft/300m	2010		Numbered Cerachrom LN	40mm	YG	Gold 116610 with black dial
Submariner Date	116619LB	3135	1000ft/300m	2010		Numbered Cerachrom, LB	40mm	WG	White gold with blue bezel and dial

OYSTER PERPETUAL SEA-DWELLER

The Sea-Dweller has an ambiguous and complex history. It debuted as a brand in 1967 after several years of development and rigorous testing.

Born of the Submariner, the Sea-Dweller is a more substantial and serious diving tool developed in collaboration with the French marine engineering and commercial diving company, Comex SA.

The early Submariner ref. 5513 provided the platform for the first Sea-Dwellers. The helium escape valve (HEV) on the side of the Oyster case distinguishes them as Sea-Dwellers. They are also uniquely numbered, with caseback engravings showing where and to whom they were issued.

A small number of the first Sea-Dwellers were distributed to retail outlets and co-branded with jewelers like Tiffany's. Only a few of these have surfaced. They are curious examples of hardcore professional tool watches sold at unlikely, high-end, refined luxury jewelers. These co-branded Sea-Dwellers command a collector's price premium and continue to be controversial.

As the Submariner grew to become the recreational scuba diver's watch, the Sea-Dweller was for the professional saturation diver. They are set apart by technical specifications such as the Sea-Dweller's enhanced caseback for improved depth rating and on later models, visual cues like the 60-minute markers around the bezel. The date complication was not a modern convenience but a useful instrument during dives and decompression. Many of these dives and subsequent decompressions lasted many days and even weeks.

Each Sea-Dweller reference had several distinct variants with a variety of dial differences, caseback engravings, and movements. All feature an HEV on the 9 o'clock side of the Oyster case. The Sea-Dweller is the only vintage Rolex to feature caseback engravings (preceding the modern Milgauss 116400), and it has more rows of dial text than any other Rolex.

Helium gas escape valve

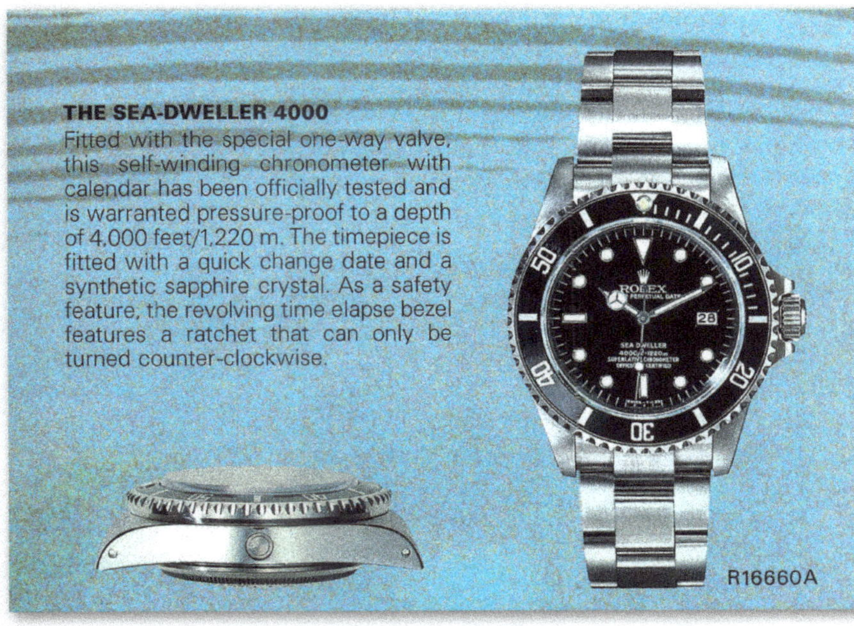

THE SEA-DWELLER 4000
Fitted with the special one-way valve, this self-winding chronometer with calendar has been officially tested and is warranted pressure-proof to a depth of 4,000 feet/1,220 m. The timepiece is fitted with a quick change date and a synthetic sapphire crystal. As a safety feature, the revolving time elapse bezel features a ratchet that can only be turned counter-clockwise.

R16660A

Extract from the Rolex Submariner complimentary booklet with each new watch

Sea-Dweller ref. 1665 'Great White'

SEA-DWELLERS VARIANTS

- **Ref.1665**
  - Double Red
  - Great White
  - Single Red
- **Ref.16660**
  - Triple 6 gloss
  - Triple 6 matte
- **Ref.116660**
  - Deepsea
  - Deepsea DeepBlue
- **Ref.16600**
  - Sea-Dweller
- **Ref.116600**
  - Sea-Dweller 4000
  - 12660
  - 126660

DOUBLE RED SEA-DWELLER REF. 1665 SPECIAL NOTE

The Double Red Sea-Dweller ref. 1665 was the first commercially available Sea-Dweller, and for collectors, it is the most coveted. Its official name is the Sea-Dweller Submariner 2000. The dial has this name printed as two rows of red text, giving it the nickname the Double Red Sea-Dweller, or DRSD for short.

The reference ran for an impressive ten years (1967 until 1977). During this run, five distinct dial variations appeared.

MK I Dial (1.6M – 2.2M): This appeared on the earliest examples before the granting of the HEV patent. MK I dials appear with a caseback engraved "Patent Pending". The MK I dial has red lettering that is a consistent size and weight. The Rolex coronet is of a thicker style, consistent with other Professional watches of the period.

MK II Dial (1.6M – 3.5M): This dial can be paired with the caseback engravings that read "Patent Pending" or "Patented". The font size and weight of the red lettering differ with the SEA-DWELLER appearing larger and heavier than the SUBMARINER 2000 text below it.

MK III Dial (2.6M – 4.0M): This dial is distinguished from the MK II by an updated coronet, which appears longer and less thick.

MK IV Dial (3.0M – 5.2M): This last version has a thicker and heavier font across both white and red lettering. It also has a pronounced coronet. This dial should be matched with a date wheel featuring closed, rather than open, 6s and 9s.

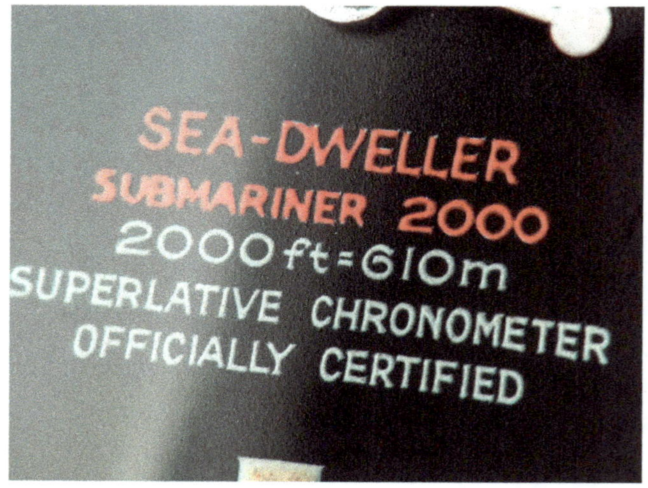

In addition to dial variations, there were two distinct styles of caseback engravings. Earliest versions of the Sea-Dweller have a caseback with "ROLEX" written horizontally across the outside surface. Following versions, beginning with the Great White in 1977, have their engravings appearing around the circumference caseback.

Stamped on the inside of the caseback are the model number and a partial serial number (last three digits). Later models have the full serial number. Some also have a stamped production date indicating quarter and year (e.g. II 72 for Q2 1972).

Example renderings of the caseback variations

ROLEX
OYSTER PERPETUAL DATE
25

ROLEX
OYSTER PERPETUAL DATE
12
SEA-DWELLER
2000 ft=610m
SUPERLATIVE CHRONOMETER
OFFICIALLY CERTIFIED

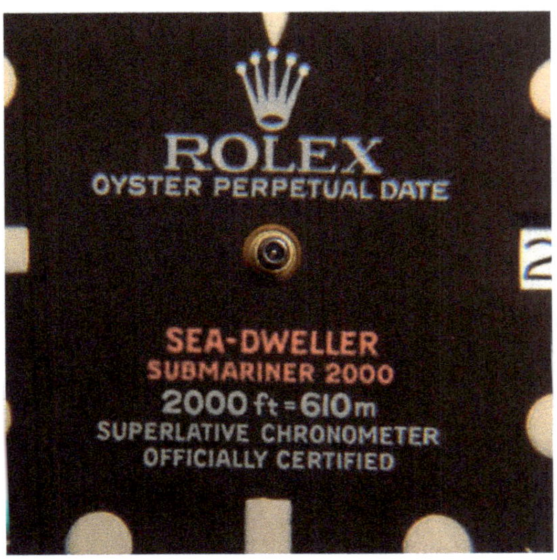
ROLEX
OYSTER PERPETUAL DATE
SEA-DWELLER
SUBMARINER 2000
2000 ft=610m
SUPERLATIVE CHRONOMETER
OFFICIALLY CERTIFIED

OYSTER PERPETUAL DATE
12
SWISS – T < 25

ROLEX
"OYSTER"
"ROLEX PATENT"
GAS ESCAPE VALVE

MONTRES ROLEX S.A.
GENEVA
SWITZERLAND
PATENTED
STAINLESS STEEL
1665

GREAT WHITE SEA-DWELLER REF. 1665 SPECIAL NOTE

In 1977 the ref. 1665 received a minor dial makeover with white lettering replacing the red. Also gone was a reference to the Submariner. By this stage, the Sea-Dweller was an established and independent product line, not just a heavy-duty Submariner.

The Great White 1665 continued in production until the early 1980s. Production overlapped with its replacement, the newer 16660 which launched in 1978. For several years, the Great White 1665 was sold alongside the new 16660 as inventory was run down.

The Sea-Dweller ref. 1665 Great White set the standard for Sea-Dweller design aesthetics until the release of ref. 126600 in 2017 (with the return of red lettering). Like its predecessor the DRSD, the Great White featured five dial variants. Frustratingly, the collector community has numbered them 0 to 4, rather than 1 to 5 like the DRSD. This choice is a result of the discovery of the earlier MK0 after the MK1.

Two examples of the MK I dial that have aged very differently

The Pisani Dial

Marcello Pisani (1956 – 2015) was a renowned Italian vintage Rolex collector, aficionado, academic, researcher, and historian.

He is largely credited with bringing COMEX to the attention of collectors. It's fitting then that the Sea-Dweller reference 1665 Mk 1 dial has come to be named after him. This variant has unique characteristics, including a 600 meters depth rating (as opposed to 610 meters) and a peculiarly shaped Rolex coronet.

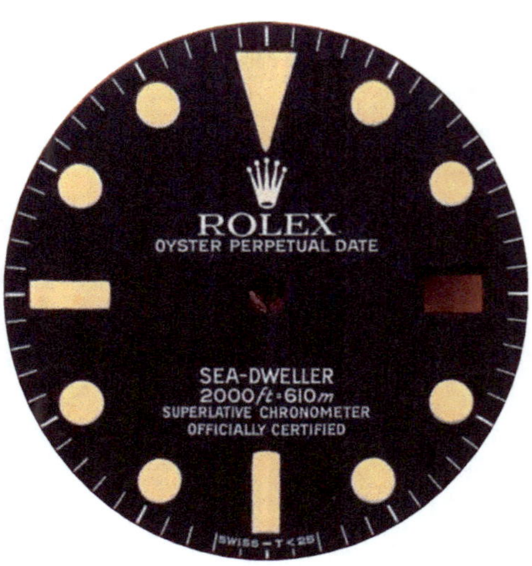

MK IV dial

MK 0 Dial: This is by far the rarest of the Great White dials and has SEA-DWELLER text that is longer and wider than the depth rating printed below it. Also of note is the closed "6" on the depth rating, 2000 ft = 600m. This closed font 6 won't match the date wheel's open font 6s.

MK I Dial (6.1M - 6.2M): The SEA-DWELLER and depth rating text, line up and appear justified, unlike on the Mk 0. Also, the "6" on the depth rating now appears closed.

MK II Dial (5.7M - 6.2M): This is a rail dial, with the "C" of Chronometer aligned with the "C" of Certified below it. The depth rating units (ft, m) now appear in italics.

MK III Dial (6M - 6.8M): Unlike the rail dial (MKII) the "C"s are not aligned, and the fonts have a different appearance.

MK IV Dial: This is hard to distinguish from the Mk III, with the difference being a differently shaped 6 in the depth rating and a smaller "R" at the end of Chronometer.

Sea-Dweller ref. 1665 dial variations from Mk0 to MK5

COMEX SEA-DWELLERS

While the following references have been co-branded with COMEX, it is still correct to have Submariners and Sea-Dwellers with HEVs and no overt COMEX dial branding. These examples will have caseback engravings indicating their special issue designation. COMEX dive watches were never sold through retail outlets and will have no other co-branding.

Submariner COMEX 5513

Submariner COMEX 5514 (154 Units)

Submariner COMEX 1680

Submariner COMEX 16800

Submariner COMEX 168000

Submariner COMEX 16610

Sea-Dweller COMEX 1665

Sea-Dweller COMEX 16600

Sea-Dweller COMEX 16660

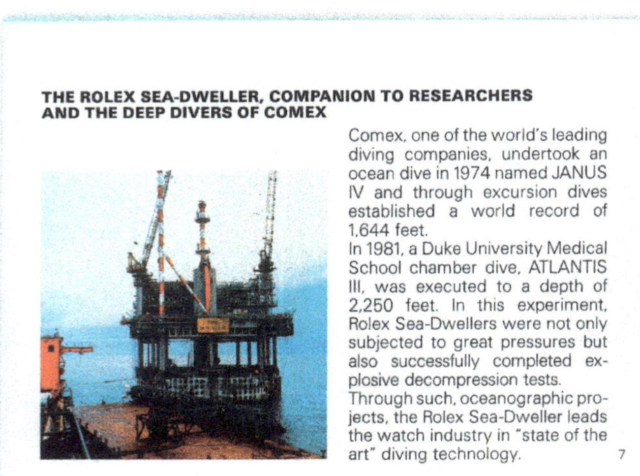

THE ROLEX SEA-DWELLER, COMPANION TO RESEARCHERS AND THE DEEP DIVERS OF COMEX

Comex, one of the world's leading diving companies, undertook an ocean dive in 1974 named JANUS IV and through excursion dives established a world record of 1,644 feet.
In 1981, a Duke University Medical School chamber dive, ATLANTIS III, was executed to a depth of 2,250 feet. In this experiment, Rolex Sea-Dwellers were not only subjected to great pressures but also successfully completed explosive decompression tests.
Through such, oceanographic projects, the Rolex Sea-Dweller leads the watch industry in "state of the art" diving technology.

7

DEEPSEA SEA-DWELLER

In 2008, Rolex introduced the Sea-Dweller DEEPSEA ref. 116660. This reference is not vintage or even modern classic, but an essential part of the Sea-Dweller story. It was a significant upgrade to the modern classic ref. 16600 and introduced the Ringlock system. This innovation sets the DEEPSEA apart from the Sea-Dweller by setting the depth rating to 12,800ft.

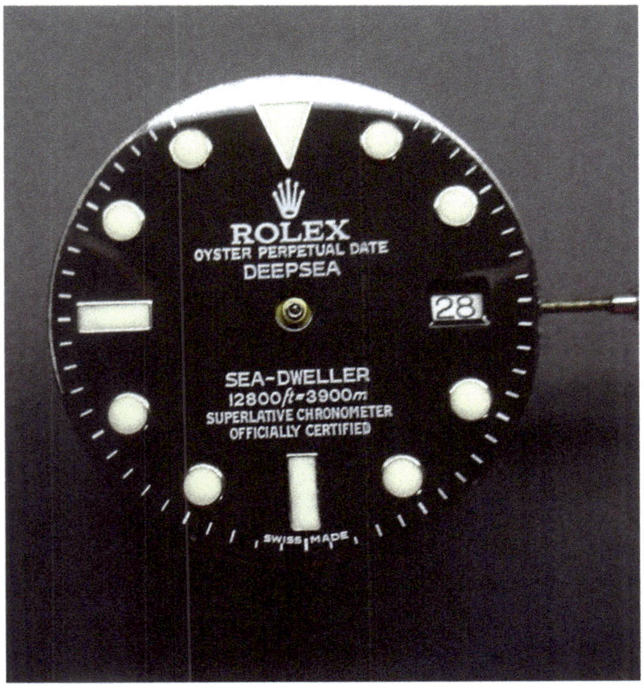

The Ringlock system is a combination of three innovations. These included a grade 5 titanium caseback capable of flexing under incredible pressure. It also had a 5mm thick sapphire crystal positioned above an inner steel ring. This Ringlock system reset the bar for professional dive watches, and as such has become a potential collectible.

DEEPSEA SEA-DWELLER DEEPBLUE JAMES CAMERON EDITION

The James Cameron Deepblue is a thoroughly contemporary watch and not even close to being considered a modern classic. However, some characteristics mark it as a sleeper, and it has the potential to be highly collectible in the future.

As far as commemorative special edition watches go, this one is unusual in that it celebrates both a record-breaking achievement and the person who performed it. Rolex commissioned it to celebrate James Cameron's successful descent to the deepest point on our planet, located at 10,900m below the surface of the Pacific Ocean. This record-setting dive was the second significant trip to the bottom of the Pacific accompanied by a special project Rolex.

The Deepsea Challenger submersible vehicle (DCV1) reached the bottom of the Mariana Trench on March 26th, 2012. It was built in Australia by the research and design company Acheron Project Pty Ltd and piloted by Canadian film director James Cameron. It carried scientific sampling equipment and high-definition 3D cameras.

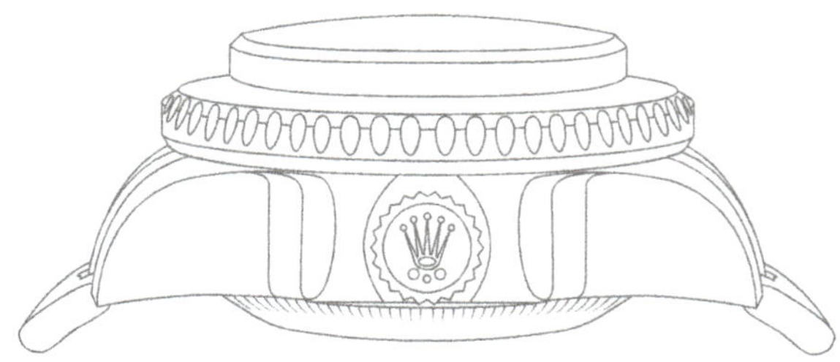

The oversized Rolex Deepsea Challenge Concept Watch was never put into production. The few prototypes are now museum exhibits.

The submersible featured a robotic arm, and attached to it was the Rolex Deepsea Challenge Concept Watch. The concept watch was a super-sized version of the standard Deepsea, created in only a few weeks and numbering only 5 or 6 units. It had a diameter of 51.4mm and a thickness of 28.5mm, 14.3mm of which was a unique sapphire crystal. The submersible and its special watch reached the ocean's deepest point after two hours and 36 minutes of descent from the surface.

The Trieste deep-diving research bathyscaphe, with its crew of Jacques Piccard and US Navy Lieutenant Don Walsh, achieved a similar feat on 23 January 1960. Reaching a depth of 10,911 meters (35,797 ft), it was accompanied by the custom-built Rolex Deepsea Special. Piccard & Walsh attached a white dial Rolex DeepSea Special Mk 2 with engraved caseback number 3 on the outside of the bathyscaphe Trieste.

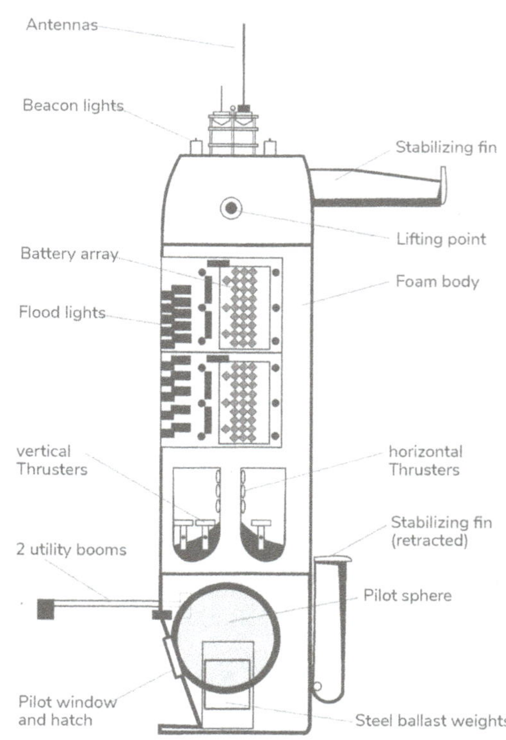

General assembly drawing of DCV1, circa 2012

The commercial Deepblue version of the concept watch has a distinctive blue and black gradient dial with vibrant green colored DEEPSEA text. This color is the same green as that of the Deepsea Challenge submersible.

Rolex designed the James Cameron Sea-Dweller Deepblue for water resistance of 12,000m. This rating is over three times that of the Deepsea.

The first available version of the Rolex Deepsea Sea-Dweller Deepblue (ref. 116660) is an homage to the concept watch. It is currently the highest performing dive watch made by Rolex. It is also the largest (17.7mm thick and 44mm diameter) with a 5mm thick crystal and a titanium caseback, featuring a patented Ringlock system. Combined, they enable massive water resistance to 3900m.

Curiously, the Deepblue shares the same reference number as the black-dialed Deepsea (ref. 116660). Critics of the Deepblue claim it is too large and top-heavy, requiring a wider redesigned Oyster bracelet. Despite the criticism, it is destined to become a future classic and a collectible reference.

The Rolex Deepsea Sea-Dweller Deepblue was revised in 2017 and replaced with ref. 126600. Rolex redesigned the Oyster case with thinned (narrower) lugs but with a broader lug width of 22mm. This design adjustment resolved the top-heavy criticisms of the 116660.

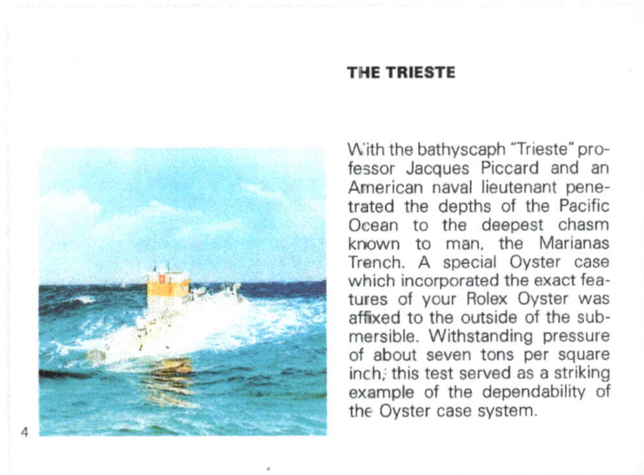

THE TRIESTE

With the bathyscaph "Trieste" professor Jacques Piccard and an American naval lieutenant penetrated the depths of the Pacific Ocean to the deepest chasm known to man, the Marianas Trench. A special Oyster case which incorporated the exact features of your Rolex Oyster was affixed to the outside of the submersible. Withstanding pressure of about seven tons per square inch, this test served as a striking example of the dependability of the Oyster case system.

4

Extract from the Rolex Submariner complimentary booklet with each new watch

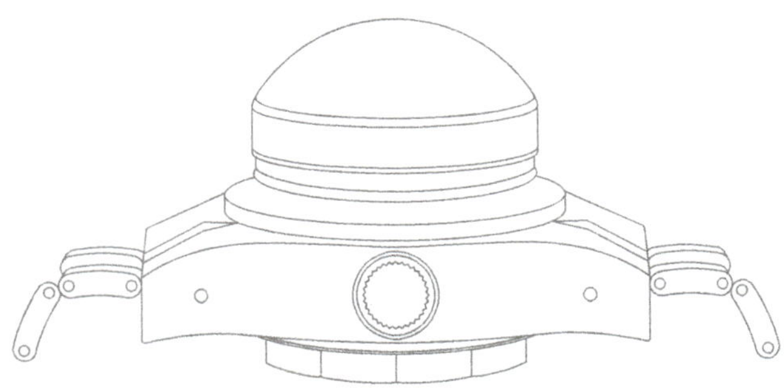

The original Rolex Deepsea Special along with several prototypes are on display at various museums

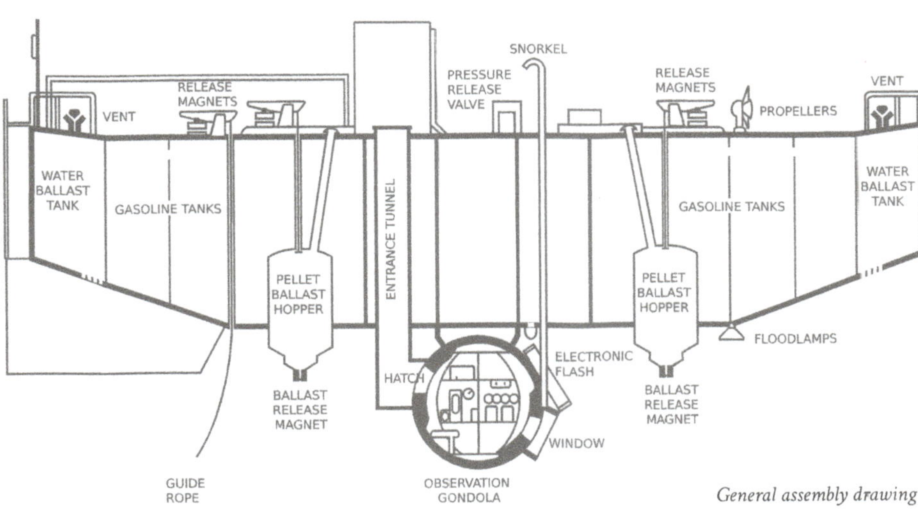

General assembly drawing of Trieste, circa 1959

SEA DWELLER REFERENCES

Name	Ref	Cal	Depth Rating	Start	End	Size	Description
Single Red	**1665**	1575	500m/1650ft	1967	1968	40mm	Prototype with only 12 examples known, derived from the Submariner 5514
Double Red	**1665**	1575	610m/2000ft	1967	1977	40mm	Included red text Submariner 2000. Inside caseback engraved with last three digits of serial number and production quarter and year
Great White	**1665**	1575	610m/2000ft	1977	1983	40mm	
Triple 6 matte	**16660**	3035	1220m/4000ft	1978	1989	40mm	Case back engraving gained an extra coronet
Triple 6 gloss	**16660**	3035	1220m/4000ft	1985	1989	40mm	Adopted white gold surrounds and dropped the hyphen between "Sea" and "Dweller" Prone to crazing/spidering. Tritium Dial (marked T<25)
Sea-Dweller	**16600**	3135	1220m/4000ft	1989	2008	40mm	1998 – 1999 Luminova dial (marked Swiss). 1999 – 2002/3 Super Luminova dial with lug holes (marked Swiss Made). 2002/3 – 2008 Super Luminova dial, no lug holes (marked Swiss Made)
Deepsea	**116660**	3135	3900m/12,800ft	2008	2018	44mm	17.6mm thick
Deepsea D-Blue	**116660**	3135	3900m/12,800ft	2014	2018	44mm	James Cameron Edition with Deepblue dial
Sea-Dweller 4000	**116600**	3135	1220m/4000ft	2014	2017	40mm	14.8mm thick
Sea-Dweller	**126600**	3235	1220m/4000ft	2017		43mm	Cyclops & red text with black dial. 50th anniversary edition
Sea-Dweller	**126660**	3235	3900m/12,800ft	2017		43mm	Cyclops & red text with black and Deepblue dial

OYSTER PERPETUAL YACHT-MASTER

While the Yacht-Master is a new and modern watch, still in production, early examples are becoming sought-after classics. The larger and more technical Yacht-Master II has a design aesthetic that divides opinion and has sold in modest volumes. These characteristics of low sales volume and polarized customer opinion are markers of future collectability. Interestingly, customers responded to the early Cosmograph Daytona and Explorer II in the same way – divided opinion – resulting in low production volume. The Yacht-Master is a series that long-term collectors should keep a close eye on.

The Yacht-Master debuted at the Basel Fair in 1992 with the reference 16628 in 18k gold, a 3135 movement, and a typical 40mm Oyster case. Subsequent iterations included Mid-Size (35mm) and Ladies Size (29mm) versions. The model received an update in 2012 with the six-digit reference 116622, and again in 2015 with the ref. 116655.

The Yacht-Master was conceived as a precious-metal, luxury take on the Submariner. It has since become a distinct product line with a specific sporting niche. Available in gold and platinum cases, they are substantial on the wrist, and the signature sunburst or platinum dials are eye-catching.

The platinum versions are proving the most popular. The hands are fatter than the Submariner's, which improves the visual balance and overall aesthetic. The red second-hand complements the red dial text, and the bi-directional bezel has a matching platinum insert with raised and polished numbers. The Oyster case is highly polished on all surfaces (no satin finish). This distinctive fully-polished look is unique to the Yacht-Master.

All Yacht-Masters use the ubiquitous Oyster bracelet, with the newest version ref. 116655 (2015), offered on a rubber Oysterflex strap. This model is the first Rolex has delivered on a rubber strap.

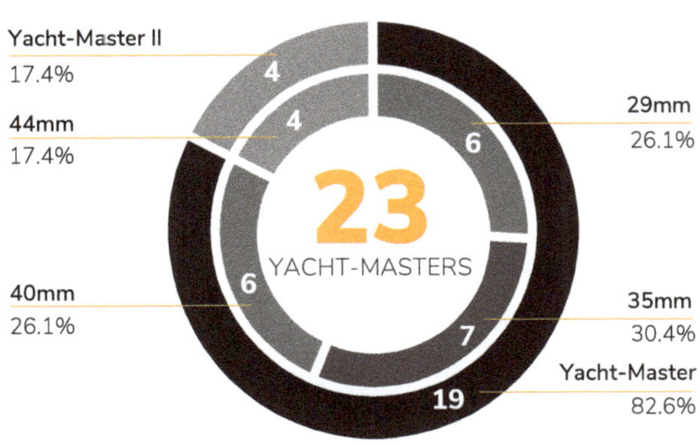

Cover of the Rolex Yacht-Master complimentary booklet with each new watch

YACHT-MASTER REGATTA II

The Rolex Yacht-Master Regatta II is a particularly polarizing design. At 44mm in diameter, it requires a muscular and tanned forearm to pull off. This model is a highly specialized niche instrument featuring a unique (to Rolex) regatta 10-minute countdown and fly-back chronograph.

The start-up sequence for this compilation is convoluted, involving a bezel twist, depressing the reset pusher, and (counterintuitively, considering it's likely to get wet) unscrewing the crown. Presumably, yachtsmen can perform this feat while staying in control of their yacht, battling the wind and waves, and still being first across the starting line!

YACHT-MASTER REFERENCES

Name	Ref	Cal	Start	Size	Case
Yacht-Master	**16628**	3135	1992	40mm	YG
Yacht-Master	**16628**	3135	1992	40mm	YG
Yacht-Master	**68623**	2135	1994	35mm	YGSS
Yacht-Master	**69623**	2135	1994	29mm	YGSS
Yacht-Master	**68628**	2135	1994	35mm	YG
Yacht-Master	**69628**	2135	1994	29mm	YG
Yacht-Master	**68628**	2235	1994	35mm	PLSS
Yacht-Master	**69682**	2235	1994	29mm	RG
Yacht-Master	**168622**	2235	1999	35mm	PLSS
Yacht-Master	**169622**	2235	1999	29mm	PLSS
Yacht-Master	**168622**	2235	1999	35mm	SS
Yacht-Master	**168623**	2235	1999	35mm	YGSS
Yacht-Master	**168628**	2235	1999	35mm	YG
Yacht-Master	**169622**	2235	1999	29mm	PL SS
Yacht-Master	**169623**	2235	1999	29mm	YGSS
Yacht-Master	**16622**	3135	1999	40mm	PLSS
Yacht-Master	**16623**	3135	2005	40mm	YGSS
Yacht-Master II	**116680**	4161	2010	44mm	SS
Yacht-Master II	**116681**	4161	2010	44mm	SSRG
Yacht-Master II	**116688**	4161	2010	44mm	YG
Yacht-Master II	**116689**	4161	2010	44mm	WG
Yacht-Master	**116622**	3135	2012	40mm	PLSS
Yacht-Master	**116655**	3135	2015	40mm	RG

YG : Yellow Gold | SS : Stainless Steel | PL : Platinum | RG : Rose Gold

OYSTER PERPETUAL EXPLORER

With such a long and rich history, the Explorer line of watches has been subject to evolution and multiple revisions. There are 14 different calibers used in approximately 24 references. Other models share many of these calibers as well, making it challenging to ensure that antique and vintage examples are original. Despite this (or because of this), Explorers are incredibly desirable and highly collectible, with prices to match.

The Rolex Explorer resulted from 20 years of research in extreme conditions. This development involved equipping climbing expeditions in the 1930s, long before the historical conquering of Everest. Rolex was one of several watchmakers who sponsored mountaineers with timing instruments and funds. The goal of this sponsorship was to create the ultimate mountaineer's tool watch capable of surviving low-pressure altitude and temperatures as low as -50C (-58F).

Contrary to popular perception, Sir Edmund Hillary did not wear an Explorer during his Everest expedition. He was wearing a Smiths watch, and a Rolex was one of two worn by Sherpa Tenzing Norgay on that first summit of Everest on May 29th, 1953. Rolex based Norgay's Explorer on an existing tough-wearing Bubbleback, modified for the expedition.

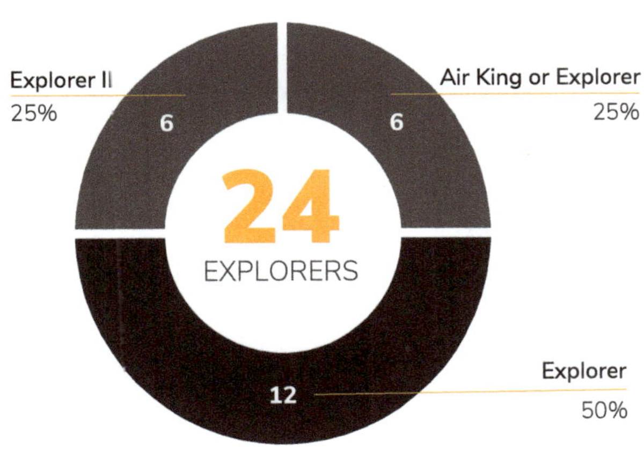

EXPLORER MOVEMENTS

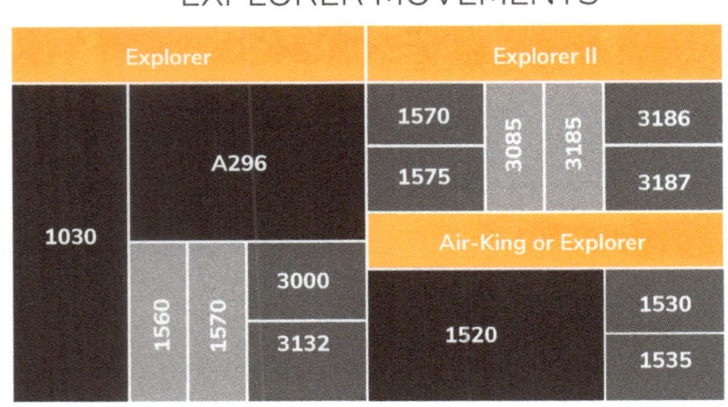

PRE-EXPLORERS

In 1952, the references 6098 and 6150 (cal. A296) provided the platform for Explorer development. They had white dials and alpha style hands, and did not bear the name Explorer yet.

A year later in 1953, these models evolved into the references 6298 and ref. 6350. Both now had the iconic Explorer 3-6-9 Arabic dials and Mercedes-style hands.

Before conquering Everest in May 1953, Rolex did not use the Explorer trademark, despite registering it in Geneva on January 26th of that same year.

REF. 6350

Rolex designed the ref. 6350 for explorers and mountaineers by building in the ability to survive significant temperature variations and the low pressure of high altitudes. It had a readable dial that performed well in poor visibility, a strengthened case, and lubricants able to tolerate temperature variations from -20C to +40C. Rolex added 'Explorer' to the dials of ref. 6350 after 1953.

EXPLORER REF. 6150 & 6610

In 1959, ref. 6610 replaced the pre-Explorer ref. 6150, on which production was ceased. These references were indistinguishable from one another, except for the new cal. 1030 movement in the ref. 6610. This new movement design allowed for a flatter caseback compared to the Bubbleback style of the ref. 6150. Collectors cite these two references as the first real Oyster Perpetual Explorers.

OYSTER PERPETUAL EXPLORER II REF. 1655

Of the four models that bear the Explorer II branding, only ref. 1665 is considered vintage. The five-digit references that followed are Modern Classics.

Rolex unveiled the Explorer II ref. 1655 in a 39mm Oyster case. In a departure from the mountaineering focus of the original Explorer, this watch was built for cave explorers (aka spelunkers or speleologists) who spent extended periods in darkness and required both a date and AM-PM indicator.

The first 1655s have a large and very distinctive, bright orange 24-hour hand. These were changed to red in later 1655s, but they still aged to orange, meaning that all 1655 24-hour hands look orange today.

The ref. 1655 was not a commercial success. Some attribute this to two shortcomings. The orange hand could not be set independently and was limited to showing AM and PM as a 24-hour clock. Also, the fixed bezel prevented the complication from working as a timer or tracking dual time, as a GMT-Master could.

The ref. 1665 is highly sought after today, even gaining unique nicknames. The first is Freccione, which is derived from the Italian word for arrow, a reference to the orange arrow-tipped hand. The second is the Steve McQueen Rolex, despite zero evidence he ever owned one.

While there is an ardent following, the 1970s dial aesthetic is an acquired taste. The 'disco-dial' as it has come to be known, is iconic to the 70s, unique among the tool watches, and instantly recognizable.

There were seven dial variations produced during the lifespan of the ref. 1655. The dial maker Stern, produced five of the variants, while Beyeler made the two service replacement dials. The service dials remained available after the watches production ended and were stockpiled to service the watches over the following decades.

This beautiful 1655 has been personalized with a tropic crystal (no cyclops) and a Jubilee bracelet

REF. 1655 DIAL VARIANTS

MK1 Dial (1971–1972)

The first iteration of the ref. 1655 had a wide coronet with a rounded foot on the "R" of Rolex. The base of the dial was marked "T SWISS T" below the 6 o'clock marker to indicate the use of tritium lume on the dial.

MK 2 Dial (1972–1977)

The second series had the coronet appearing more splayed, with more substantial dots on the points of the coronet. Nicknamed the 'Frog Foot', this style of coronet also appears on Explorer ref. 1016.

Another difference between the MK 1 and MK 2 Explorer II dials is the spacing of the word "PERPETUAL" which is narrower on the MK 2. The top of the letter 'L' lines up with the bottom right stroke of the letter 'X' in "ROLEX" (on the MK I, the 'A' of "PERPETUAL" sits below the 'X'). The base of the dial is marked "T SWISS T" below the 6 o'clock marker, indicating tritium lume on the dial.

MK 3 Dial (1974–1977)

This version is what collectors term a 'Rail Dial'. The name refers to the alignment of the text, "SUPERLATIVE CHRONOMETER OFFICIALLY CERTIFIED" on the bottom half of the dial. The letter Cs in words "CHRONOMETER" and "CERTIFIED" are aligned, leaving a distinctive vertical gap in the middle of the two lines of text. Similar Rail Dials also appear on some ref. 1665 Sea-Dwellers. The coronet differs from the MK 2 by being better defined and more symmetrical. It is marked "T SWISS T" on the outer bottom edge.

MK 4 Dial (1977–1980)

The coronet on the MK4 is narrower and taller than the MK 3, and the text alignment of the lower dial text is the same as the MK 2. The most noticeable change is the text at the bottom edge of the dial, which now reads "T SWISS < 25 T" in a serif font.

MK 5 Dial (1979–1984)

The last of the production dials also had the "T SWISS < 25 T" at the bottom edge of the dial, but without serifs on the lettering (sans-serif font). The coronet is narrower than the MK 4 and has a larger, more open oval at the bottom. This coronet is very similar to those on the Maxi-dial Submariner ref. 5513.

MK 6 & MK 7 Dials (from 1984)

The MK 6 service dial was the last to use tritium "T SWISS <25 T" and the MK 7 was the first to use Luminova "SWISS".

BEZEL VARIATIONS

There are four bezel variations plus a service version. The differences are in the font size (weight) and the alignment of numerals.

MK 1 Bezel (1971–1973)

These appeared with a thick, bold font and the base of the numerals sit close to the inner edge of the bezel. The MK 1 bezel is correct for both MK 1 and MK 2 dials.

MK 2 Bezel (1973–1977)

These used a similar thick bold font with numerals aligned around the center of the bezel ring, rather than sitting on the inner edge. These bezels have appeared on some MK1 dial watches but mostly on watches with MK 3 and MK 4 dials.

MK 3 Bezel (1977–1980)

These used a thinner font with numerals positioned centrally on the bezel. The MK 3 bezel is correct for MK 3 and MK 4-dialed watches.

MK 4 Bezel (after 1980)

The numerals on this bezel use a thin font, like the MK 3 bezel. A noticeable difference is a long hook on the 1s. The MK 4 bezel is correct for MK 5-dialed watches.

Extract from the Rolex Explorer complimentary booklet with each new watch

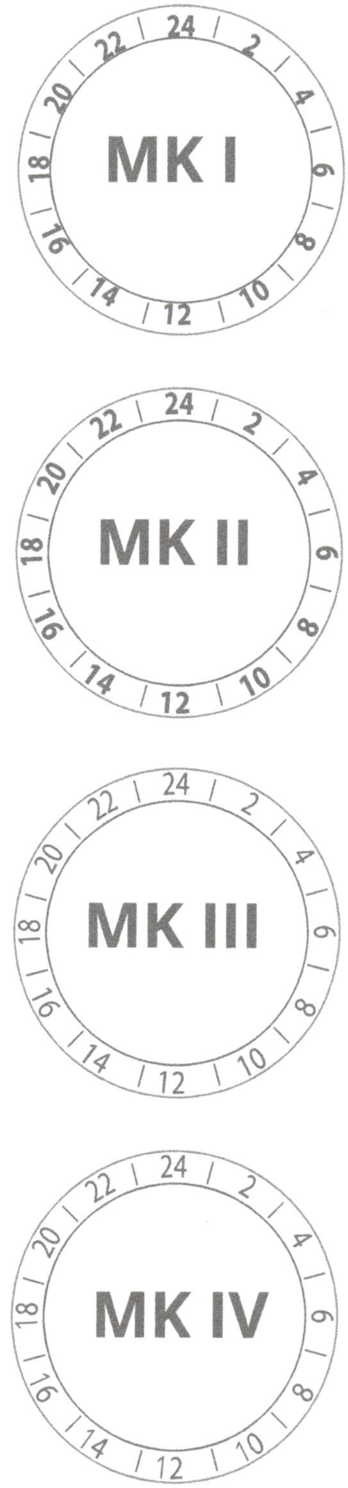

Explorer II ref. 1655 bezels. MK 1 to MK 4 plus a service bezel

REF. 16550 (1985)

Rolex introduced the ref. 16550 with a larger 40mm Oyster case and a sapphire crystal. Ref. 16550 was a transitional reference in all respects and was offered in black and white Rail Dials. It had white gold surrounds to the hour markers and adopted the GMT-Master-style GMT hands.

The white dial version has come to be known as the 'Polar Explorer', and a paint defect has turned them a pleasing cream color today. Polar Explorers command a considerable price premium over the black dial variants. Rolex corrected this paint defect before the end of the production run and introduced it along with changes to hour markers. These changed from white gold to black.

These revised dials may well have been brought forward into the production of ref. 16550 from its pending successor, the ref. 16750.

The new movement shared with the GMT-Master, now allowed the GMT hand to be independently set, enabling dual time zones.

REF. 16570 (1989)

Cosmetically, this reference differed from its predecessor by the black color of the hour marker surrounds. The main difference justifying the new reference number was an upgraded movement, cal. 3185. Subsequent revisions of this reference saw enhancements like Luminova, a laser-etched rehaut and the no-holes case (elimination of drilled lugs).

REF. 216570

This modern version arrived in an enlarged 42mm Oyster case with another upgraded movement, cal. 3187. The new dial adopted maxi-style hour markers and Chromalight lume, but the most obvious visual change was the return of the orange, 'Big Triangle'-style GMT hand, reminiscent of the ref. 1665.

Ref. 16570 (1989)

Ref. 216570

EXPLORER REFERENCES

Name	Ref	Cal	Start	End	Bezel	Size	Case	Description
Air-King or Explorer	**5507**	1520			Smooth (Polished)	34mm	SS	Quickset date
Air-King or Explorer	**5504**	1530			Smooth (Polished)	34mm	SS	Super Precision
Air-King or Explorer	**5701**	1535			Smooth (Polished)	34mm	SS	Perpetual Chronometer
Air-King or Explorer	**5501**	1520	1953		Smooth (Polished)	34mm	SS	Precision
Air-King or Explorer	**5500**	1520	1957		Smooth (Polished)	34mm	SS	Mercedes & lollipop hands
Air-King or Explorer	**5507**	1520	1958		Smooth (Polished)	34mm	SS	Quickset date
Explorer	**8044**	1030			Smooth (Polished)	36mm	SS	Perpetual Chronometer
Explorer	**8045**	1030			Smooth (Polished)	36mm	GF	Perpetual Chronometer, Golds Shell
Explorer	**14270**	3000			Smooth (Polished)	36mm	SS	Perpetual Chronometer
Explorer	**114270**	3132			Smooth (Polished)	39mm	SS	
Explorer	**6150**	A296	1952	1959	Smooth (Polished)	36mm	SS	Bubble back
Explorer	**6298**	1030	1953		Smooth (Polished)	36mm	SS	Perpetual, Precision
Explorer	**6298**	A296	1953		Smooth (Polished)	36mm	SS	Perpetual Chronometer
Explorer	**6299**	A296	1953		Smooth (Polished)	36mm	SSYG	Perpetual Chronometer
Explorer	**6350**	A296	1953		Smooth (Polished)	36mm	SS	Dials feature "Explorer"
Explorer	**6610**	1030	1959		Smooth (Polished)	36mm	SS	36mm
Explorer	**1016**	1560	1963	1989	Smooth (Polished)	36mm	SS	Perpetual Chronometer
Explorer	**1016**	1570	1963	1989	Smooth (Polished)	36mm	SS	Perpetual Chronometer
Explorer II	**1655**	1570	1971	1981	Numbered	39mm	SS	Big orange 24-hour arrow hand
Explorer II	**1655**	1575	1971	1981	Numbered	39mm	SS	Big orange 24-hour arrow hand
Explorer II	**16550**	3085	1985	1989	Numbered	40mm	SS	Red and black GMT arrow hand
Explorer II	**16570**	3185	1989	2012	Numbered	40mm	SS	Red and black GMT arrow hand
Explorer II	**16570**	3186	1989	2012	Numbered	40mm	SS	Red and black GMT arrow hand
Explorer II	**216570**	3187	2012		Numbered	42mm	SS	Anniversary big orange GMT arrow hand

SS : Stainless Steel | GF : Gold Fill | YG : Yellow Gold

OYSTER PERPETUAL MILGAUSS

The Milgauss is a rather obscure model that remains somewhat below the radar. In recent years, it has gained more traction with collectors and prices for vintage references in any condition are rising.

The earliest examples are easily mistaken for an Oyster Perpetual. Contemporary examples are less prone to this, thanks to some unique design cues.

In 1956, Rolex introduced the Oyster Perpetual Milgauss, ref. 6451, intended for scientists and engineers working in power plants, medical facilities, and research labs. The name comes from the Latin mille, meaning one-thousand, and gauss, the unit of measure for magnetism.

The watch was developed and tested in the 1950s by the European Organization for Nuclear Research (CERN), the world's pre-eminent particle physics laboratory. The design was a technological breakthrough solving a real problem for timekeeping in these electrically and magnetically polluted environments.

Reference 6541 was very similar in appearance to the Rolex Submariner, which had proved to be a winning design. The movement was enclosed in a Faraday cage of ferromagnetic alloy, making it resistant to magnetism up to 1000 gauss.

The original Milgauss sold in modest volume, presumably because this was a small professional niche in the 1950s. Today they are rare and sought after by collectors. In stainless steel only, they are discreet and low-key, yet instantly recognizable by vintage enthusiasts.

The Milgauss has a short history compared to other Professional models. Only two updates were released, ref. 6541 and a modified 1019, in the 1960s and 1970s. The updates introduced a silver dial option. Some 1019s were also released without the unique lightning bolt second hand, though many had it added later during service.

For 20 years consumers largely ignored the Milgauss. Then in 2007, Rolex revived their "scientist's watch", releasing three new references starting with model ref. 116400.

Due to the internal magnetic shield, the Milgauss is thicker than other Professional watches, including the similarly designed Submariner. As with all Rolex Oyster cases, the Milgauss is waterproof up to 100 meters (330 feet).

A 50th-anniversary model features a black dial with a distinctive green-tinted sapphire crystal. This Glace Verte (GV) model is named for its green tinted glass and is the only Rolex to feature it. The GV is also the only sapphire crystal without a Laser Etched Coronet (LEC), as the green tint makes it too hard to see, even under magnification.

MILGAUSS REFERENCES

Name	Ref	Cal	Start	End	Bezel	Size	Description
Milgauss	**6543**	1080	1953	1956	Numbered, LN	38mm	Only 80-200 thought to exist. 19mm Lugs
Milgauss	**6541**	1080	1956	1960	Numbered, LN	38mm	Introduced the lightning bolt second hand and bigger 20mm lugs. Includes a full ferromagnetic Faraday cage
Milgauss	**6019**		1960	1963	Polished (Smooth)	38mm	Silver or black matte dial, straight second hand and smooth bezel. A silver CERN dial exists with no luminous markers.
Milgauss	**1019**	1580	1963	1988	Polished (Smooth)	38mm	CERN dial variant also available
Milgauss	**116400**	3131	2007		Polished (Smooth)	40mm	Black, white and blue (2014) dial variants
50th Anniversary Milgauss	**116400GV**	3131	2016		Polished (Smooth)	40mm	50th Anniversary green tinted crystal. Does not have the LEC

LN : Lunette Noir (black)

COSMOGRAPH DAYTONA

A chronograph is a stopwatch integrated with a conventional time-of-day watch, and Rolex has been making these since the 1930s. Rolex chronographs have evolved through three distinct eras: 1) the four-digit reference Valjoux era; 2) the five-digit reference Zenith era; and 3) the contemporary six-digit reference, in-house era.

Rolex offered the now typical two pusher configuration starting in 1937 with Valjoux movements. These are all based on numerous variants of the caliber 72. In the 1980s, Rolex adopted the Zenith El Primero movement in the 16000 series models and in 2000, Rolex debuted their in-house caliber 4130 in their six-digit model references.

Although we commonly associate the name Cosmograph with the pre-Daytona, it was the triple-date Moonphase dress watch that used it first. Cosmos is a reference to the night sky illustrated on the moon phase subdial.

The vintage Cosmograph and early Daytonas are extremely technical to service and highly challenging references to collect. With countless dial variations and numerous reference iterations, they make the other tool watches straightforward by comparison.

Headline-making auction prices continue to attract both admirers and nefarious dealers. Most collectors now acknowledge that the majority of pre-Daytonas in circulation do not have entirely original dials. An underground cottage industry has been tampering with dial text for several decades in the pursuit of higher prices. In particular, aftermarket red text is being added to dials, creating an abundance of Daytona 'Big Reds'. Writing is also being removed to create the "Rolex"-only, Solo Dials. Experts refrain from public judgment, as most cannot be definitively confirmed, and Rolex maintains their customary silence on the matter.

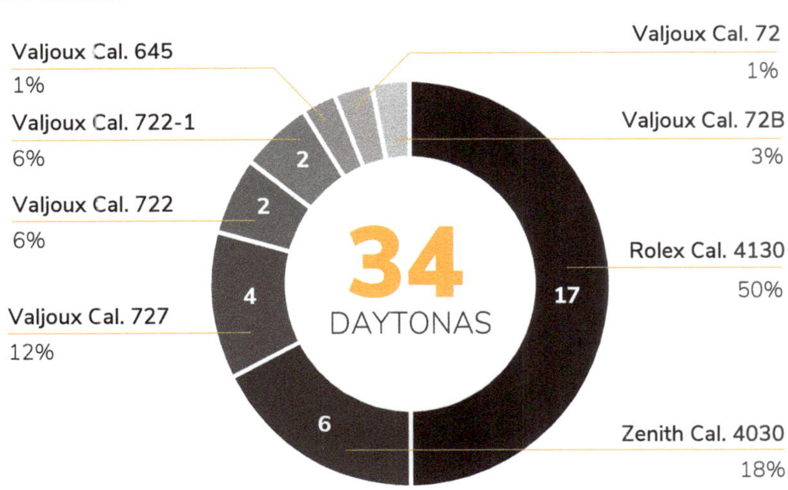

Cosmograph Daytona movements

EXOTIC CHRONOGRAPH DIALS

Rolex made chronographs in relatively small numbers compared to other models, and considerable effort was invested in making them appealing to mainstream buyers. One of these strategies included small runs of unusual dial designs, many of which were used as market tests.

THE MAYER DIAL

This was named after the recording artist, guitarist, and watch collector, John C. Mayer. This is an exotic Daytona dial with three colors and the words "Rolex Oyster Cosmograph". This dial is black in color with white subdials and a distinctive red outer minute track (i.e. 3 colors). This black and white combination is most often associated with Oyster cases having screw-down pushers like reference 6240.

This is a controversial dial with many enthusiasts doubting authenticity. The dial was never featured in an official Rolex catalog, which typically bestows bona fide credentials on a watch. The manufacturers' stamp "SINGER" on the back is unusual and not typical or consistent with their other dials supplied to Rolex during this period. The dials do not appear before the 1990s and only in the USA. These factors are sufficient to raise doubts but are not definitive.

Mayer dial Cosmographs have successfully been serviced by official Rolex Service Centers (RSCs), however, they have been known to service watches that have subsequently been proved to have highly convincing fake dials. Even RSCs can be fooled by a high-quality counterfeit dial, as have prestigious auction houses.

Christie's in Geneva successfully auctioned a Mayer Cosmograph (Lot 187) on Nov 12, 2007 for US$160,000. This and other auction examples are cited as convincing votes of legitimacy. The collector community is still divided, with passionately held views on both sides for and against.

THE PATRIZZI DIAL

A Patrizzi dial appears on the Rolex Daytona reference 16520. It features subdials that have oxidized and turned brown. The imperfection only appears in some, but not all examples.

The effect is the result of the inconsistent formulation of an organic varnish called Zapon.

Zapon lacquer is a cellulose nitrate varnish patented in 1887. It is a solvent mixture containing amyl acetate, butyl acetate, and propyl acetate. Zapon lacquer dries to a very thin transparent film that was originally sold as a varnish for polished metal parts, a fixative for watercolor paintings, and a consolidant for manuscripts degraded by iron gall inks. It has since known to turn yellow with age.

Some Daytona examples with Zapon varnish have aged prematurely, turning brown and tinting the silver subdial beneath. These color changes are progressive and do not stabilize over time, making each Patrizzi dial unique. While this uniqueness makes them desirable among collectors, it has also caught the attention of unscrupulous dial-bakers (artificially aged through heat exposure) and counterfeiters. Fake Patrizzi dials are increasingly common.

The dial was named after Mr. Osvaldo Patrizzi, an Italian auctioneer. He discovered the effect in 2005 while evaluating a collection of ref 16520s for auction – all featuring unique brown subdials. While the imperfection is considered a defect, they are nonetheless beautiful and command a considerable price premium.

THE INVERTED 6 DIAL

Some examples of the Daytona 16520 have a curious inverted six on the hour subdial. The number '6' is oriented upside down and reads as a '9'. Later models have the number 6 in the normal orientation. The inverted 6 is synonymous with the Mk 1 dials described later in the 16500 series.

PRE-COSMOGRAPH

The ref. 6234 was launched in 1955 and did not yet bear the Cosmograph name. This model was an evolution of the Dato-Compax chronographs of the antique period. It had both a tachymeter and a telemeter scale printed on the outer edge of the dial, under the crystal. Tachymeters measure speed and telemeters distance, both equally important to race car drivers and downhill skiers (see Antique section, Dato-Compax). Both instruments used either imperial (mph) or metric (kph) measures, but not both.

The ref. 6234 was made at the rate of about 500 per year until production ended in 1961. Ref. 6234 was not a commercial success. However, the roughly 3000 units made during the seven-year production run makes them a scarce and desirable reference today.

COSMOGRAPH DAYTONA

Dial printing on the first Cosmograph Daytona (ref. 6239 in 1963) was evolutionary. Beginning as a Rolex Cosmograph, the Daytona name was added the following year (1964). These first Cosmograph references appeared with double Swiss dials, double Swiss underline dials, and double T-Swiss-T dials. There is no generally accepted explanation for these duplicate markings. Each variation appears within a narrow range of serial numbers.

The screw-down pushers arrived in the 1970s with ref. 6263, introducing waterproofing. These pushers were in response to owners ruining their watches by trying to run the chronograph under water. This addition initially claimed to waterproof the watch to 50m/165ft, but by the 1980s, was uprated to 100m/330ft.

An unusual feature of the 6263 is that it used higher grade versions of the Valjoux 23 movement in precious metal cases. The 18K gold cases had chronometer grade movements, while the stainless steel versions did not. This practice continued until the use of "Superlative Chronometer, Officially Certified" started to appear on dials in the 1970s.

Valjoux movements continued to power Cosmograph Daytonas until 1987, when Rolex switched suppliers to Zenith and their automatic El Primero movement. Renamed cal. 4030, this powered the Daytona ref. 16520 until 1987.

By this stage, the name had evolved from Cosmograph Daytona to Daytona Cosmograph to just Daytona.

Rolex switched to their in-house cal. 4130, with the launch of ref. 116520 in 2000. After 80 years, this was Rolex's first in-house chronograph.

The cal. 4130 required changes to the subdial configuration and placement of the minute and hour counters moved to the left and right positions. The seconds counter moved from the left side of the dial to the bottom, and the two horizontal subdials shifted to seven degrees above the center point. Some speculate this was an anti-counterfeiting measure rather than a technical engineering requirement.

Vintage and Classic Daytonas have been the subject of extensive research, with numerous books written about them in recent years. These three are particularly noteworthy:

- Rolex Daytona Manual Winding by Osvaldo Patrizzi and Guido Mondani (ISBN 978-88-940669-9-9)
- Rolex Daytona Story by Osvaldo Patrizzi and Guido Mondani (AISN B009HQ87PS)
- The Ultimate Rolex Daytona - Miniature edition by Pino Abbrescia, Fabio Santinelli, Paolo Gobbi and Naomi Ornstein

COSMOGRAPH REF. 6265

COSMOGRAPH REF. 6239

Rolex introduced the first Cosmograph ref. 6239 in 1963. It had a modest 36.5mm diameter and did not yet officially bear the name Daytona. It acquired the nickname thanks to Rolex becoming the official timekeeper at the Daytona Raceway in Florida.

The ref. 6239 had been redesigned for race car drivers and found informal celebrity endorsement with Paul Newman. This ref. 6239 has come to be known as the Paul Newman Daytona and was the first reference to use inverse colors for subdials. This color scheme has come to be known as a 'panda dial', with the inverse color combination being the 'reverse-panda dial'. All previous chronograph dials are monochrome.

Like the first of most new models, it has several variations in dials, bezels, and movements. Valjoux 72B, 722 and 722-1 are all correct for the ref. 6239.

This Cosmograph was the first chronograph to have the tachymeter scale moved from under the glass on the dial to outside the watch on the bezel. Singer made the dials and stamped the backs with "Singer Brevets AV".

The first ref. 6239s had serial numbers starting around 923,000, and they featured a satin-finish bezel. These uniquely calibrated tachymeter scales are known as the '275 Intermediate'. The intermediate-style tachymeter reads 300, 275, 250, 255, and so on.

With the next iteration around 1964, Rolex removed the 275 indicator along with the small dash or hash markers, replacing them with dots or pips, which are easier to read. Also gone was the satin brushed finish, replaced with a mirror-polish finish.

Dials were marked "SWISS" only and accompanied by an underline below the top dial text. This small horizontal line is thought to indicate the use of tritium lume, before the formal adoption of "T SWISS T" markings in 1964.

The name Daytona appears on dials below the text, "Rolex Cosmograph". These are often described as floating logos, as the Daytona name has an offset vertical spacing with the Cosmograph text above, and appears to float in space. Floating logos disappear by 1966.

Rolex scholars estimate total production ref. 6239 at about 105,000 units.

COSMOGRAPH DAYTONA REF. 6241

The 6241 is closely related to its predecessor, the 6239. It is distinguished from the ref. 6240 by the pump pushers. The subsequent 6240 screw-down pushers were to significantly change the character and water resistance of the series.

The 6241 is a high-ticket vintage chronograph in stainless steel, but a solid gold version like this example is an extremely rare and precious watch. The yellow gold examples were made in both 14K and 18K gold and destined for different territories.

Produced between 1965 and 1969, the ref. 6241 was made in small numbers and the solid gold version especially so. The main differentiator between earlier references is the introduction of a bakelite bezel insert – a striking departure from the stainless steel bezels.

DAYTONA REF. 16500 SERIES

Running from 1988 to 2000, this modern classic Daytona is increasingly sought after by collectors. It features the high-performance Zenith caliber 400 (El Primero), which has been heavily customized by Rolex for use in this series of chronographs. Two hundred distinct modifications were made before becoming known as the Rolex caliber 4030, a Superlative Chronometer Officially Certified (SCOC) watch. With modern screw down crown and pushers, it is waterproof to 100m, like the other tool watches in the Professional collection.

While some owners claim the dial is busy and hard to read, it is a practical watch suitable for daily wear. Prices for this reference vary considerably based on the six reference variations, with the stainless steel ref. 16520 being the most desirable.

The first versions appear on R-serial numbers from 1987 to late 1988. These were made in small volume and are quite rare. Their dials usually feature the floating Cosmograph with "Officially Certified Chronograph" text, and the bezel is calibrated to "200, 180, 160", etc. To complicate matters, exceptions have been seen with intermediate bezels showing the 225 indicator from the Cosmograph predecessors. These unusual combinations are considered official as they've appeared in Rolex marketing materials. This Daytona was offered on a 78360 bracelet that's fully brushed, with no polished links.

The second version appears on L-serial numbers from 1989 to early 1990. It differs from the first R-serial series with a bezel calibrated 400, 300, 250, 220, etc. These dials do not have "Officially Certified Chronometer" text on them. These were made in small numbers and very rare.

The third versions appear on E, X, and N-serial numbers from early 1990 to late 1991. They differ from the second version by dial markings only. These have "Superlative Chronometer Officially Certified" text and an unusual inverted digit 6 in the lower subdial. Numerals are printed with a more rounded font style and the minute track has shorter markers.

The fourth version appears on N, C, and S-serial numbers from late 1991 to early 1993. While the dial is the same as the third series, it was available on a new bracelet ref. 78390 with polished center links. This version is thought to have been produced in larger volumes than the third series. With a larger supply on the pre-owned market, prices are lower than earlier versions.

The fifth versions appear on S, W, T, and U-serial numbers from early 1993 to late 1998. This version was made in higher volume and its iterations are not rare. It features a typical tritium dial with the digit 6 on the subdial now being the right way up. The minute track markers appear fatter and more prominent.

The sixth and final series on U, A, and early P-serial numbers are from late 1998 to early 2000. The only difference being the Luminova dial.

DIAL VARIANTS OF THE 16520

MK 1: This dial appears on Daytona's with R (1988) and L (1989) serial numbers and appears to have only been in production for about one year. It is a tritium dial and labeled "T SWISS T" at the six o'clock position.

It features the word "Cosmograph" positioned closer to the center of the dial, (aka the "Floating Cosmograph"). Less visually obvious is the fact that the whole text "ROLEX, Oyster Perpetual, Superlative Chronometer" is positioned lower than later dials.

The subdials all have a rounded and smaller font. The minute subdial at the three o'clock position has four minute markings between the five-minute markers. Later dials have only three of these "dash" marks. The hour subdial (or hour totalizer) may have the inverted six into early 1995 serial numbers.

MK 2: This dial was also only produced for about a year appearing in the early 16520s. The main difference is that the words "Officially Certified" are omitted despite the watch being rated as a chronometer.

The word "Cosmograph" is now positioned higher and grouped more tightly with the rest of the text at the top of the dial. This is also a tritium dial but labeled "T SWISS MADE T" at the six o'clock position.

MK 3: This dial is very similar to the current production dials, with the words "OFFICIALLY CERTIFIED" reintroduced again. The hour totalizer subdial may still have an inverted 6.

MK 4: This is the most commonly used dial used on the 16520 and most typically associated with the Daytona. The dial text is tightly grouped and situated slightly higher than the original dials.

The subdial fonts have evolved to become bolder, more square, and subsequently more legible. With fewer hash marks on the subdials, the overall effect is cleaner and less busy.

MK 5: This is essentially the same dial as the Mark 4, with the exception of the luminous material. Daytonas in the serial range beginning A (1999), until the end of production in 2000 (serial range P), have Luminova used in the markers and hands. These dials are subsequently labeled "SWISS" at the six o'clock position.

There are acknowledged transitional examples in the U-serial range from late 1998 that have Luminova dials and SEL bracelets. These are considered transitional, but otherwise correct.

COSMOGRAPH DAYTONA REFERENCES

Ref	Cal	Depth Rating	Start	End	Bezel
6238	72B	0	1960	1976	Polished (smooth)
6239	72	0	1963	1969	
6239	722	0	1963	1969	
6239	722-1	0	1963	1969	
6240	722	50m	1965	1969	
6241	645	0	1969	1970	
6241	722-1	0	1968	1970	
6262	727	0	1970	1970	
6263	727	100m/330ft	1971	1987	
6264	727	100m/330ft	1970	1980	
6265	727	100m/330ft	1970	1980	
6269	727	100m/330ft	1980	1990	Diamond
6270	727	100m/330ft	1980	1990	Diamond
16515	4030	100m/330ft	1991	1999	
16518	4030	100m/330ft	1991	1999	
16519	4030	100m/330ft	1991	1999	
16520	4030	100m/330ft	1991	1999	
16523	4030	100m/330ft	1991	1999	
16528	4030	100m/330ft	1991	1999	
116503	4130	100m/330ft	2000	2016	
116505	4130	100m/330ft	2000	2016	
116508	4130	100m/330ft	2000	2016	
116509	4130	100m/330ft	2000	2016	
116515	4130	100m/330ft	2000	2016	
116518	4130	100m/330ft	2000	2016	
116519	4130	100m/330ft	2000	2016	
116520	4130	100m/330ft	2000	2016	
116523	4130	100m/330ft	2000	2016	
116528	4130	100m/330ft	2000	2016	
116589	4130	100m/330ft	2000	2016	
116598	4130	100m/330ft	2000	2016	
116599	4130	100m/330ft	2000	2016	
116500LN	4130	100m/330ft	2016		
116515LN	4130	100m/330ft	2016		
116518LN	4130	100m/330ft	2016		
116519LN	4130	100m/330ft	2016		

OYSTER PERPETUAL AIR-KING

The Air-King brand has a long aviation pedigree and is one of the longest-running product lines in Rolex history. The first Air-King appeared in 1945 and arrived as a conscious effort to merge predecessors featuring names like Air-Lion, Air-Tiger, and Air-Giant. These monikers had been in use on various Bubbleback Rolex Oyster Perpetuals since the early 1930s. They became popular among WWII pilots of the British Royal Air Force since they proved more reliable, accurate, and readable than the RAF-issued timepieces.

Rolex reused references across model lines. Because of this, it is correct for a ref. 5500 to appear in both product lines with either Air-King or Explorer dials. These dual-reference models are challenging to validate with the conversion from one to the other, which is a matter of a simple dial swap. For example, a prized 3-6-9 Explorer dial fitted to a less desirable Air-King 5443 can net an unscrupulous dealer a tidy profit.

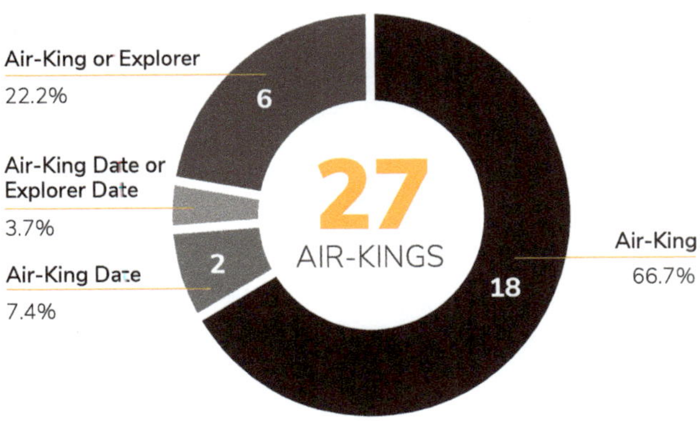

Air-King reference variations

AIR-KING MOVEMENTS

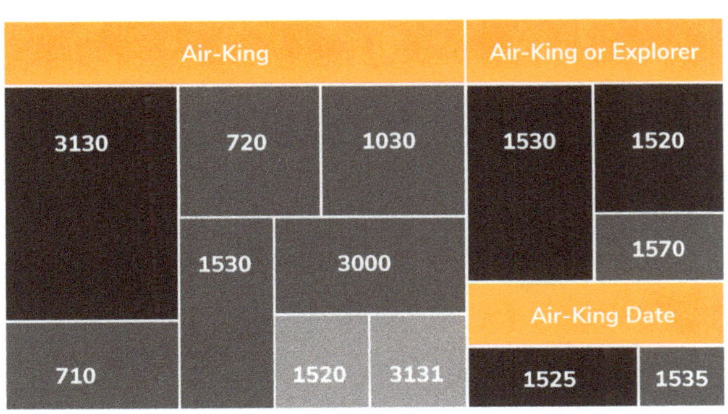

Air-King movement calibers

There's a lot to learn and love about the Air-King. Broadly speaking (and there are exceptions), the 4000 series reference numbers are WWII era watches from the 1940s. The 6500 series appear in the early 1950s. The 5000 series references appear in the mid-1950s and run until the late 1970s. The line was dormant during the 1980s before reappearing again in the early 1990s as a five-digit reference (14000 series).

The first Air-King references used manual wind Aegler Hunter movements (like cal. A720, A296). These included ref. 4925, 4365 (also a dual-reference) and 4499. In the 1950s Rolex released Air-Kings with what are considered Rolex in-house movements like the cal. 1030, 3000, 1520, 1530, 3130 and the modern cal. 3131. The first auto-winding, Oyster Perpetual Air-King, which was ref. 6552 with cal. 1030, arriving in 1953.

The iconic Air-King ref. 5500 appeared in 1957 (with cal. 1530 and 1520) and lasted 37 years. It is one of the longest-running references in the Rolex lineup. The ref. 5500 is a robust and reliable watch that can be serviced by any competent, independent watchmaker. With supply still plentiful, prices are reasonable, making this a popular first vintage Rolex choice for those new to collecting the brand.

Confusingly (and perhaps infuriatingly), the ref. 5500 was introduced in 1957, then again with the same reference number in 1970, but with an upgraded chronometer movement (cal. 1570).

The Air-King line of watches used manual winding movements, but adopted automatic self-winding movements (Perpetuals) in the 1950s. Some of these Perpetuals featured a date complication and were branded "Air-King Date". These automatics had thicker Oyster case dimensions.

The manually wound Air-Kings used Precision and Super Precision-class movements, which are described in Chapter 11. These references are considered desirable for their slimmer case profiles.

Through the 1950s and 1960s, the Air-King formula remained unchanged. The 34mm (Mid Size) Oyster complemented the larger Datejust (36mm). During the 1970s, the Air-King line received movement upgrades to the chronometer cal. 1570, and eventually achieved full COSC certification in 2007.

Rolex retired the Air-King in 2014. By this time it had become a thoroughly contemporary watch and was offered in 26mm, 31mm, 34mm, and 36mm sizes. The decision to retire the Air-King was to make way for the Rolex Oyster Perpetual and avoid the two lines competing in the same market segment.

The Air-King was relaunched with a radical and modern reinterpretation in 2016 with ref. 116900. This new interpretation is also a highly polarizing design, giving it the hallmarks of a future classic collectible. The dial is considered by some to be too fussy with too many colors, lacking in Rolex character, and not sufficiently reminiscent of earlier Air-Kings. Others however, like the contemporary and distinctive design with its reference to modern flight instruments.

AIR-KING REFERENCES

Name	Ref	Cal	Depth Rating	Start	Bezel	Size	Case	Description
Air-King	**1401**	710	100m/330ft	1943	Engine-Turned	34mm	SS	Precision
Air-King	**4365**	710	100m/330ft	1945	Smooth (Polished)	34mm	SS	Pre-Air-King, Air-Giant, Precision
Air-King	**4499**	720	100m/330ft	1946	Smooth (Polished)	34mm	SS	
Air-King	**4925**	720	100m/330ft	1945	Smooth (Polished)	34mm	SS	
Air-King or Explorer	**5500**	1520	100m/330ft	1957	Smooth (Polished)	35mm	SS	Batton hands, Precision
Air-King or Explorer	**5500**	1530	100m/330ft	1957	Smooth (Polished)	35mm	SS	Super Precision
Air-King or Explorer	**5500**	1570	100m/330ft	1970	Smooth (Polished)	34mm	SS	COSC
Air-King or Explorer	**5501**	1520	100m/330ft	1958	Fluted	35mm	SSYG	Precision
Air-King or Explorer	**5501**	1530	100m/330ft	1958	Fluted	35mm	SSYG	Super Precision
Air-King	**5502**	1530	100m/330ft	1958	Smooth (Polished)	35mm	GF	Precision
Air-King or Explorer	**5504**	1530	100m/330ft	1958	Smooth (Polished)	35mm	SS	Super Precision
Air-King	**5506**	1530	100m/330ft	1958	Smooth (Polished)	34mm	GF	Super Precision
Air-King	**5520**	1520	100m/330ft	1974	Smooth (Polished)	34mm	GF	Precision
Air-King Date	**5520**	1525	100m/330ft	1974	Smooth (Polished)	34mm	GF	Quickset date
Air-King Date	**5700**	1525	100m/330ft	1958	Smooth (Polished)	34mm	SS	Precision
Air-King Date or Explorer	**5701**	1535	100m/330ft	1958	Fluted	34mm	SSYG	Precision
Air-King	**6552**	1030	100m/330ft	1953	Smooth (Polished)	34mm	SS	First automatic Air-King, Precision
Air-King	**6652**	1030	100m/330ft	1953	Smooth (Polished)	34mm	SS	Transitional to the 5500
Air-King	**14000**	3000	100m/330ft	1989	Smooth (Polished)	34mm	SS	COSC
Air-King	**14010**	3000	100m/330ft	1989	Engine-Turned	34mm	SS	COSC
Air-King	**114200**	3130	100m/330ft	2007	Smooth (Polished)	40mm	SS	COSC
Air-King	**114210**	3130	100m/330ft	2007	Engine-Turned	40mm	SS	Precision
Air-King	**114234**	3130	100m/330ft	2007	Fluted	40mm	WG	COSC
Air-King	**116900**	3131	100m/330ft	2017	Smooth (Polished)	40mm	SS	COSC
Air-King	**14000M**	3130	100m/330ft	2000	Smooth (Polished)	34mm	SS	COSC
Air-King	**14010M**	3130	100m/330ft	2000	Engine-Turned	34mm	SS	COSC

SS : Stainless Steel | YG : Yellow Gold | GF : Gold Fill | WG : White Gold

OYSTER PERPETUAL GMT-MASTER

This iconic reference boasts an incredible history marked by technological innovation. Many subtle design variations add to their mystique, making this reference enjoyable to learn about and challenging to collect.

The GMT-Master has well-documented connections to early jet aviation and pioneering space flight. The James Bond 007 (Honor Blackman) and Magnum PI (Tom Selleck) associations only add to the romance.

All versions of the GMT feature a bi-directional bezel. These were friction fit (non-clicking) until the arrival of the five-digit references, with their 120 click bezel. It is correct for transitional models like the 16750 to have the older-style friction fit, non-clicking bezels.

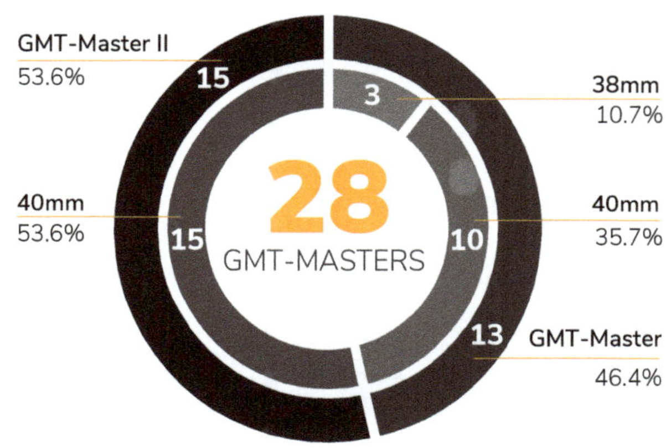

GMT-Master references

GMT-MASTER MOVEMENTS

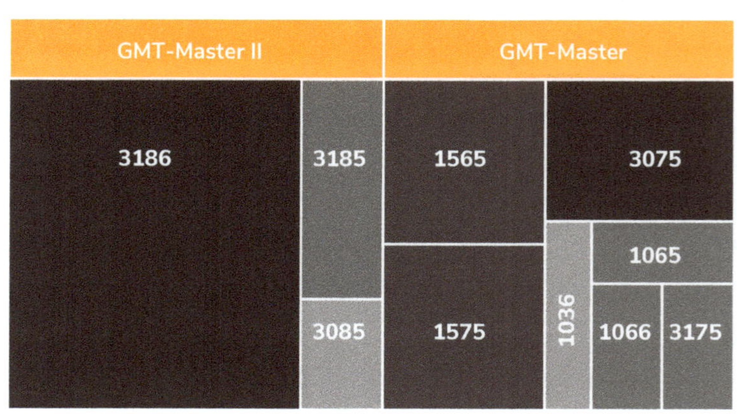

GMT-Master Calibers

REF. 6542 (1954–1959)

Believe it or not, Fidel Castro liked his Rolex watches and his GMTs in particular. Many photographs show him wearing two different Rolex on the same wrist. He set one to local time in Havana and the other displayed the time in Moscow.

Castro frequently gave Rolex watches to his commanders and associates. These were essential equipment for field commanders and crucial for coordinating and synchronizing military operations. In a pre-quartz era, the precision of a Rolex tool watch was an obvious choice.

The first GMT arrived with a gilt dial and radium Bakelite bezel, but without the crown guards. The bezel used a fragile Bakelite (acrylic plastic) insert, which was prone to cracking and was only available in red and blue. This first reference was famously worn by Pussy Galore (Honor Blackman) in the James Bond classic, Goldfinger.

In 1959, the ref. 6542 was subject to an embarrassing recall due to the high levels of radioactivity from the radium in the Bakelite bezels. The US Atomic Energy Agency demanded the recall, securing the ref. 6542's place in GMT folklore. A few examples that went unreturned are still in existence and command significant collector premiums. Even examples with replaced bezels are rare and sell for high prices.

Collectors of these original radium-filled bakelite bezels largely shrug off any health risks associated with their all-original 6542s. Most cite the fact these are so precious as to be rarely worn and that radiation levels are relatively low. They will also make light of the fact their other vices pose bigger health risks!

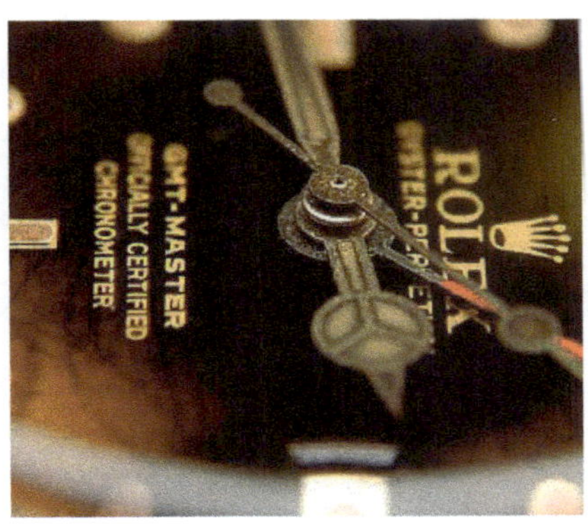

REF 1675 (1959–1979)

Ernesto "Che" Guevara appears in many pictures with his GMT-Master ref. 1675, which was a gift from a fellow communist leader, Fidel Castro.

Guevara was killed on October 9th, 1967 by CIA operative, Félix Rodríguez, who removed Guevara's GMT-Master from the wrist of his corpse, and has it in his possession to this day.

The updated ref. 1675 introduced pointed crown guards (PCG) designed to protect the crown from knocks. This case was known as the 'Cornine case', which is Italian for little antlers. It was later updated and featured the rounded ones that became standard.

The ref. 1675 was the first Rolex sports watch available with a choice of either the Oyster (ref. 7206, 7836, 78360) or Jubilee (ref. 6251, 62510) bracelets.

There were six main matte dials from MK 0 to MK 5. Dials MK 1, MK 2, and MK 5 are thought to have sub-variants. The earliest dials used gilt, closed chapter, ring dials. These featured a painted ring enclosing the outer minute track (the chapter ring).

In 1966, this model was updated with a matte dial finish and printed lettering. The first gilt dial examples sell for a significant premium over the later white-letter matte dial versions.

Before 1970, the 24hr-hand had a small triangle-shaped arrow tip. This style of arrow hand lasted into the early white-letter matte dial period. After 1970 it switched to the larger, arrow style used today.

Ref 1675 dial variations MK 0, MK 1 long E, MK 2

Ref 1675 dial variations MK 3 mini radial dial, MK 4 Maxi, MK 5

Rolex launched the ref. 1675 with a red and blue bezel insert (BLRO, bleu rouge, Pepsi). A black option (LN, lunette noire) was introduced late in the model's life cycle. The bezel is friction fitted and does not click.

This reference was COSC certified and was the first GMT-Master to use the text, "Superlative Chronometer, Officially Certified". Variations include a two-tone ref. 1675/3 and a solid 18K gold ref. 1675/8. These featured unique nipple-style index markers and an applied gold coronet (made of gold and fixed to the dial like the hour markers). The applied coronet is unique to the four-digit 1675/3 and 1675/8. With the arrival of the five-digit reference, Rolex replaced the applied crown with a gilt version, matching the dial text.

Aluminum Bezel Inserts

The colored aluminum bezel inserts for the GMT-Master 1675 have a distinctive red color on the rear (underside) of the insert.

The same bezel inserts for the GMT-Master 16750 are blue on the underside. This change is thought to have occurred with the introduction of the five digit GMT in the early 1980s.

So, red-backed inserts are for the ref. 1675, and blue for later 16750s.

Both inserts share the same dimensions and are interchangeable.

Outside diameter: 37.75mm

Inside diameter: 30.2mm

The dimensions of bezel inserts changed and became larger with later references (16700, 16760, 16710),

Outside diameter: 37.65mm

Inside diameter: 30.7mm

Convincing fakes like this white dial GMT-Master 6542 fuel speculation and debate about the authenticity of the mythical white-GMT. These counterfeits change hands for a few hundred dollars among replica-builders.

OFFICIALLY CERTIFIED

GMT-MASTER

ROLEX
24

MONTRES ROLEX S.A.
GENEVA
SWITZERLAND
PATENTED
STAINLESS STEEL
1675

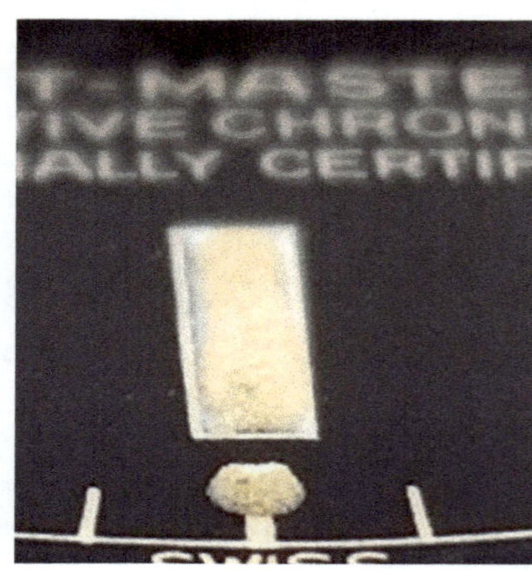

REF. 16750 (1979–1988)

Ref. 16750 GMT-Master with glossy dial

This reference arrived with movement caliber 3075 and introduced a convenient quickset date and hacking feature. Due to the new movement, the dial and hands are not backward compatible with ref. 1675. The hand stack (or stacking order) changed too, moving the hour hand to the bottom of the pile, unlike ref. 1675, where the red 24-hour hand is at the bottom. Like the earlier ref. 1675, the GMT-hand was fixed and could not be independently set. Reading second time zones required the bezel to be rotated to align with the 24-hour hand. The bezel assembly and colors were the same as ref. 1675.

The dials were matte finished with tritium lume until about 1983. Rolex then switched to the modern gloss finish with white gold index surrounds. The first versions of these glossy dials omitted the word "Date", reading only "Oyster Perpetual".

Rolex offered black or brown dials on the two-tone and solid gold references (16753 & 16758). These dials were the last to feature the nipple-style indexes and the first to have the gilt coronet (as opposed to an applied coronet). Rolex offered the ref. 16750 on either Oyster or Jubilee bracelets.

REF. 16760 (1983–1987)

Reference 16760 is the first GMT-Master II, and it launched with caliber 3085. This reference was the first GMT-Master to feature a sapphire crystal. Production overlapped with its predecessor, the ref. 16750. The new cal. 3085 extended the hacking and quickset date to include quickset of the 24-hour hand as well. This improvement allowed calculation of a third time zone. Collectors dubbed this reference the 'Fat Lady', or the 'Sophia Loren', thanks to the thicker and fuller Oyster case needed to accommodate the new movement.

Rolex made this reference only in stainless steel, and the bezel was available in two colors – black (LN) or black and red bezel, (LNRO, lunette noir rouge, Coke). There was no blue and red option (BLRO). The colored bezel insert is interchangeable with that of the GMT II 16710 and the GMT 16700.

REF. 16700 (1988–1998)

The 16700 launched with caliber 3175. Dials featured tritium lume (with the corresponding T<25 markings) until about 1998, when Luminova was adopted. Although the 16700 was the replacement for the 16750, it did not have a quickset 24-hour hand.

Rolex offered the ref. 16700 in stainless steel only but with a choice of two bezel colors – red and blue (BLRO, bleu rouge, Pepsi) or black (LN, lunette noir).

The date wheel had open 6s and 9s like its predecessors, until about 1992, when the font changed to the modern closed style.

REF. 16710 (1989–2007)

This reference features caliber 3185. It came with tritium dials (and the T<25 dial markings) until about 1997, then switched to Luminova (and the "SWISS" dial markings) in 1998, and then Super Luminova (with "SWISS MADE") in 2000.

Curiously, there was a short run of Luminova dials marked with the tritium indicator T<25, produced in 1998. This contradiction (tritium markings and Luminova lume) can be mistaken by uninformed collectors for a relumed or fake dial. Informed collectors may recognize one of these as being incorrectly and under priced.

The ref. 16710 is functionally the same as the reference it replaced ref. 16760, except it had a slimmer Oyster case. In 2000, Rolex introduced Solid End Links (SELs) to eliminate the notorious bracelet rattle. In 2003, Rolex marked crystals with a Laser Etched Crown (LEC).

Stick Dial

Conventional dial

In 2007, caliber 3186 was introduced around Z & M serial number range. Examples of the transitional ref. 16710 with the cal. 3186 movements are rare and sought after. In the same year, the case was redesigned to eliminate visible spring bar holes in the lugs. The bracelets for these SEL models had notched end links for removing the spring bars, and became permanently attached to the bracelet.

In 2005, starting in the D-serial range, the dial font evolved to a non-serif font giving rise to the 'Stick Dial'. The term refers to the Roman numerals in GMT-Master II. With the serif gone they resemble two parallel sticks. These dials lasted through the Z and M serial range. Rolex discontinued the reference in 2008.

The 16710 was available in stainless steel, two-tone (16713), and solid gold (16718), with all three bezel options (Black LN, Coke LNRO & Pepsi BLRO).

A very common fake bezel known as the 'hooked-two', for a GMT-Master II reference 16710. The hook of the "2" extends beyond the horizontal base leg

THE GMT-MASTER II

The GMT-Master II is available in steel, steel and 18 kt gold, or entirely in 18 kt gold. It is an Officially Certified Chronometer, and it is fitted with shoulders to protect the winding crown, a synthetic sapphire crystal and the following devices: instantaneous calendar, "24-hour" rotatable bezel with corresponding hand with the triangular tip, independent "12-hour" hand and a stop second hand. The GMT-Master II continues to display the correct time even while the date and time zone are being reset. The GMT-Master II Chronometer is warranted pressure-proof to a depth of 330 feet (100 m).

4

R16710 A

Extract from the Rolex Submariner complimentary booklet with each new watch

GMT MASTER REFERENCES

Name	Ref	Cal	Depth Rating	Start	End	Bezel	Size	Case	Description
GMT-Master	**6542**	1036 (1954–1959)	50m/165ft	1954	1959	Numbered, Bakelite, BLRO	38mm	SS	First GMT derived from the Turn-O-Graph (6202) featured a Bakelite bezel, no crown guards. Worn by Pussy Galore (Honor Blackman) in Goldfinger 1964
GMT-Master	**6542**	1065 (1957–1959)	50m/165ft	1954	1959	Numbered, Bakelite, BLRO	38mm	SS	
GMT-Master	**6542**	1066 (1957–1959)	50m/165ft	1954	1959	Numbered, Bakelite, BLRO	38mm	SS	
GMT-Master	**1675/3**	1565 (1959–1965)	50m/165ft	1959	1979	Numbered, BLRO, LN	40mm	SSYG	Two-tone version of the 1675 featuring black or brown nipple dial with applied coronet
GMT-Master	**1675**	1565 (1959–1965)	50m/165ft	1959	1979	Numbered, BLRO, LN	40mm	SS	First Rolex sports watch available on both the Oyster (ref 78360) and Jubilee (ref 62510) bracelet. Prior to 1970 the 24hr hand had a small triangle arrow. After 1970, it switched to the large
GMT-Master	**1675/8**	1565 (1959–1965)	50m/165ft	1959	1979	Numbered, BLRO, LN	40mm	YG	Solid gold version of the 1675 featuring black or brown nipple dial with applied coronet
GMT-Master	**1675/3**	1575 (1965–1980)	50m/165ft	1959	1979	Numbered, BLRO, LN	40mm	SSYG	Two-tone version of the 1675 featuring black or brown nipple dial with applied coronet
GMT-Master	**1675**	1575 (1965–1980)	50m/165ft	1959	1979	Numbered, BLRO, LN	40mm	SS	First Rolex sports watch available on both the Oyster (ref 78360) and Jubilee (ref 62510) bracelet. Prior to 1970 the 24hr hand had a small triangle arrow After 1970 it switched to the large
GMT-Master	**1675/8**	1575 (1965–1980)	50m/165ft	1959	1979	Numbered, BLRO, LN	40mm	YG	Solid gold version of the 1675 featuring black or brown nipple dial with applied coronet
GMT-Master	**16750**	3075	100m/330ft	1979	1988	Numbered	40mm	SS	Early versions omited "Date" from the dial. New movement introduced hand stack change 1986 saw change to glossy dial with white gold surrounds
GMT-Master	**16753**	3075	100m/330ft	1979	1988	Numbered	40mm	SSYG	Two-tone version of the 16750 (Tigers Eye)
GMT-Master	**16758**	3075	100m/330ft	1979	1988	Numbered	40mm	YG	Solid gold version of the 16750
GMT-Master	**16700**	3175	100m/330ft	1988	1998	Numbered, BLRO, LN	40mm	SS	Tritium up to 1997 and Luminova from 1998 Date wheel had open 6s and 9s until 1992/3 Available in Pepsi and Black bezel only

Name	Ref	Cal	Depth Rating	Start	End	Bezel	Size	Case	Description
GMT-Master II	**16760**	3085	100m/330ft	1983	1987	Numbered, LNRO	40mm	SS	Fat Lady, Sophia Loren. First GMT-Master II with thicker case, sapphire crystal and white gold indexes Coke & black bezel only
GMT-Master II	**16710**	3186	100m/330ft	1989	2007	Numbered, BLRO, LNRO, LN	40mm	SS	Initially tritium then switching to Luminova in 1998. Case and movement revised in 2007
GMT-Master II	**16713**	3186	100m/330ft	1989	2007	Numbered, LN	40mm	SSYG	Two-tone version of 16710
GMT-Master II	**16718**	3186	100m/330ft	1989	2007	Numbered, LN	40mm	YG	Solid gold version of 16710
GMT-Master II	**16710**	3185 (1988–1992)	100m/330ft	1989	2007	Numbered, BLRO, LNRO, LN	40mm	SS	Initially tritium then switching to Luminova in 1998. Case and movement revised in 2007
GMT-Master II	**16713**	3185 (1988–1992)	100m/330ft	1989	2007	Numbered, LN	40mm	SSYG	Two-tone version of 16710
GMT-Master II	**16718**	3185 (1988–1992)	100m/330ft	1989	2007	Numbered, LN	40mm	YG	Solid gold version of 16710
GMT-Master II	**116710**	3186	1000ft/300m	2008		Numbered, Cerachrom	40mm	SS	Introduction of maxi case, ceramic bezel and rehaut engraving
GMT-Master II	**116713**	3186	1000ft/300m	2008		Numbered, Cerachrom	40mm	SSYG	Two-tone version of 116710
GMT-Master II	**116718**	3186	1000ft/300m	2008		Numbered, Cerachrom	40mm	YG	Solid gold version of 116710
GMT-Master II	**116719**	3186	1000ft/300m	2014		Numbered, Cerachrom	40mm	WG	White gold Pepsi (BLRO)
GMT-Master II	**116758**	3186	1000ft/300m	2014		Numbered, Cerachrom	40mm	YG	Jeweled version
GMT-Master II	**116759**	3186	1000ft/300m	2014		Numbered, Cerachrom	40mm	WG	Jeweled version
GMT-Master II	**116760**	3186	1000ft/300m	2014		Numbered, Cerachrom	40mm	SS	Jeweled version
GMT-Master II	**126710**	3186	1000ft/300m	2017		Numbered, Cerachrom, BLRO, LN	40mm	SS	

SS : Stainless Steel | YG : Yellow Gold | WG : White Gold

Patina | ‘pa-tə-nəz |

noun ***fine arts***

- a fine coating on the surface of a metal.
- a surface appearance of something grown beautiful especially with age or use.
- the surface produced by accretion, a surface texture or color which works of decorative art gain through the action of time.

OYSTER WATCH Co
GENEVA SWISS
GRANT BRITAIN PATENTS
260554/1925
274.708 274.789
SWISS PATENTS
+ 114.940 + 120.848
+ 120.851 + 122.110
FRENCH PATENTS
630.179 630.180
U.S.A PATENT 1.661.232
GERMAN PATENT 443.386

1928

Chapter

08

CLASSICS & CROSSOVERS

The Rolex Classic Collection is neither formal and dressy (like the Cellini collection) nor professional and sporting. They are versatile crossover watches that are robust and equally appropriate to pair with formal wear as they are with sportswear. They share the waterproof and accuracy specifications of the Professional collection, but also feature jewel-set bezels and precious metals.

The brand trademarks in this category are some of the oldest in the Rolex stable, spanning antique, vintage, and classic genres. Surviving examples are plentiful, making them attractively priced and an excellent entry point into vintage Rolex ownership.

For most collectors, the Rolex Oyster is all about the world-famous, waterproof Oyster case, with its rounded mid case, integrated lugs, screw-down crown and proprietary threaded caseback. In common vintage parlance, the term and trademark Oyster refers to the case style. Few know that the Rolex Oyster was also used to brand a collection of significant watches before the development of the iconic Oyster case.

"Fashion passes, style remains"
Coco Chanel (1883-1971)

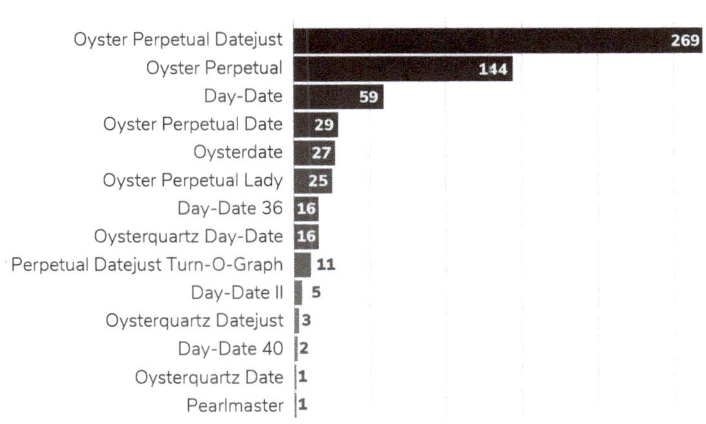

Classic References

OYSTER CASE

Cultural norms today make no distinction between men's and women's watch sizes. Today, women proudly wear men's size Rolexes, but neither men nor women have yet to show much interest in women's antique and vintage Rolexes. There is virtually no collector interest in Ladies-sized Oysters, and the market is very soft. Ladies-size Oyster watches currently sell for scrap metal and gem value.

However, there is growing appreciation among male collectors for smaller watches. 36mm down to 34mm Oyster cases wear well on smaller wrists. They are particularly comfortable if you have a pronounced wrist bone. Combined with clever design cues like thin bezels, tapered hands, and indexes, these smaller cases are said to wear and feel larger than their actual size. For antique and vintage examples produced up to the mid-1950s, 33mm to 34mm is considered a Men's Size and was quite large at the time. There are four classes of Oyster case:

Full Size: 36mm to 40mm +

Mid Size: 34mm

Boys Size: 30-32 mm

Ladies Size: 30mm or less.

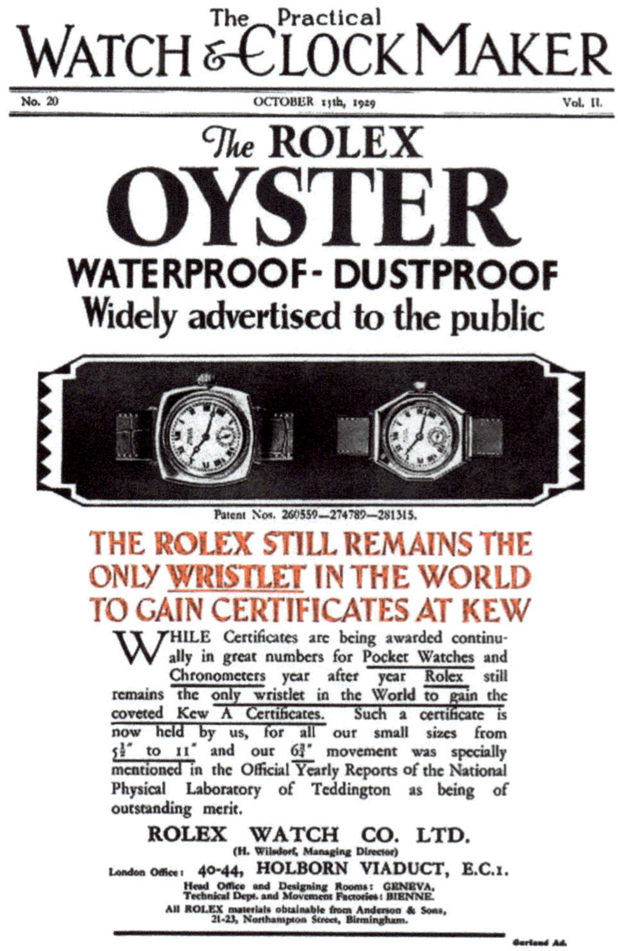
The Practical WATCH & CLOCK MAKER

No. 20 OCTOBER 15th, 1929 Vol. II.

The ROLEX

OYSTER

WATERPROOF - DUSTPROOF

Widely advertised to the public

Patent Nos. 260559—274789—281315.

THE ROLEX STILL REMAINS THE ONLY WRISTLET IN THE WORLD TO GAIN CERTIFICATES AT KEW

WHILE Certificates are being awarded continually in great numbers for Pocket Watches and Chronometers year after year Rolex still remains the only wristlet in the World to gain the coveted Kew A Certificates. Such a certificate is now held by us, for all our small sizes from $5\frac{1}{2}''$ to $11''$ and our $6\frac{3}{4}''$ movement was specially mentioned in the Official Yearly Reports of the National Physical Laboratory of Teddington as being of outstanding merit.

ROLEX WATCH CO. LTD.

(H. Wilsdorf, Managing Director)

London Office: 40-44, HOLBORN VIADUCT, E.C.1.

Head Office and Designing Rooms: GENEVA.
Technical Dept. and Movement Factories: BIENNE.

All ROLEX materials obtainable from Anderson & Sons, 21-23, Northampton Street, Birmingham.

Garland Ad.

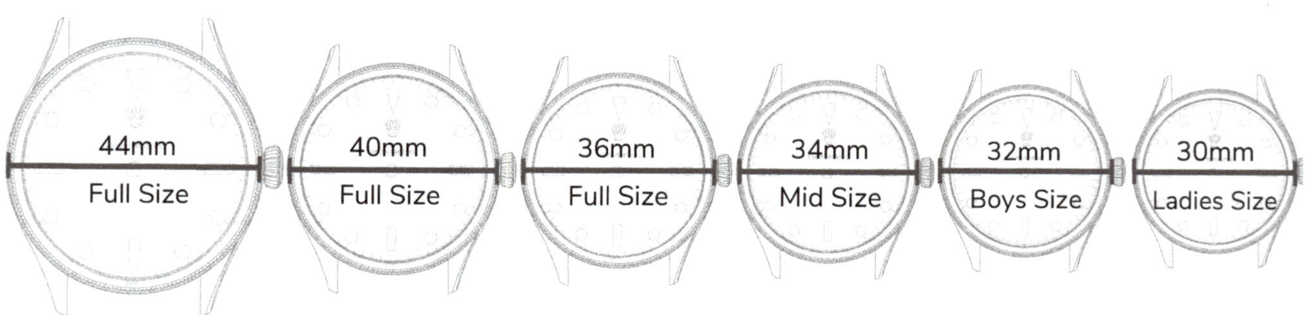

ROLEX OYSTER WATCHES

In 1926, the Rolex Oyster models made up a collection. These early models fall in the antique rather than the vintage category by virtue of their date of introduction (pre- WWII).

Initially, there were four models, including an octagonal and a cushion-shaped case. Each was available in gents (32mm) and ladies (28mm) sizes, in yellow and white gold.

In 1927, Mercedes Gleitze wore a 28mm ladies, octagonal Rolex Oyster around her neck during her famous 35km cross-channel swim attempts. She was a Rolex brand ambassador featuring in early advertising and marketing promotions.

It was a 36mm Men's cushion case on the wrist of Sir Malcolm Campbell when he broke his land speed record for the 9th and final time in 1935.

The octagonal Oyster proved less popular than the cushion case, and Rolex dropped it from the catalog after a few years. Rarity now makes them the more desirable model among collectors.

Rolex adopted the simple, two-piece Oyster case in the 1930s. They continued to improve the Oyster design with the inventions of the Twinlock and Triplock crowns in 1953 and 1970. It was these innovations that provided the platform for the success of the future sport and professional models.

The term Oyster is a registered trademark and Rolex guarantees all Rolex Oysters to be waterproof to a depth of at least 50m. This guarantee is extended to 100m for the Explorer, to 300m for the Submariner, and 1220m for the Sea-Dweller.

The original Rolex Oyster collection is therefore limited to these antique examples bearing only these words "Rolex Oyster" on the dial.

Testing the Napier Railton Bluebird at Pendine Sands, Wales. Circa 1927

Sir Malcolm Campbell sporting his cushion case Rolex Oyster

OYSTER PRECISION

Rolex watch dials featuring the text, Precision, Extra Precision, or Super Precision, are not product lines but a classification of the accuracy of the movement. A variety of case styles in addition to the Oyster contains Precision grade movements.

The terms Precision, Extra Precision, or Super Precision are successors to the earlier Prima, Extra Prima, and Ultra Prima classifications, which appeared on Bubblebacks and pocket watches.

The Rolex Oyster Precision ref. 6422 with waffle-texture dial and Explorer-style hour markers. The Oyster case shows signs of service polish

Caliber 1215 with unevenly aged service components (ratchet & crown wheels). The regulator is set to the minimum position, suggesting the movement is running fast and may be ready for a service.

Contemporary Oyster Perpetual with bronze sunburst dial and broad baton-style hands and markers.

These movements were not submitted for independent certification but were tested in-house for reasons known only to Rolex. These are not COSC certified movements.

Rolex has a long history of using independent testing labs, reflected by their use of the word "Chronometer" on early (antique) dials. Rolex changed the term in the 1930s to "Officially Certified Chronometer" which appears on early Bubblebacks. In the 1950s this became "Superlative Chronometer".

The Controle Officiel Suisse des Chronometres (COSC) was formed in 1973, displacing the Kew Observatory as the pre-eminent testing authority. COSC was subsequently adopted by Rolex as their official testing authority. Certified movements now had dials featuring the words, "Superlative Chronometer Officially Certified".

Like the early pre-Oysters, Rolex awarded the Ultra Prima or Super Precision designation to the top 10% most consistent and accurate movements. The remaining 90% were assigned either Extra Prima (or Extra Precision) or Prima (or Precision).

Prima markings appear on pre-Oyster and Bubbleback cases. Precision markings appear on Oyster cases with movements derived from the base caliber 1200.

These non-COSC (not chronometer grade) movements have a reputation for being lower quality but this is unfounded. Seasoned watchmakers will attest to the accuracy these movements can attain. They can be made to achieve accuracy within COSC standards (-4 / +6 secs per day) and even meet Rolex's own target of -1/+5 secs per day.

These manual wound Precision movements were sold alongside COSC-certified Perpetual (automatic) chronometers until they were phased out in the 1980s. Mechanically simple and very robust, Rolex produced the 1200 series from early 1954 up until late 1984.

Cal 1210 18,000 bps 1954 to 1964

Cal 1215 18,000 bps with Date 1954 to 1964

Cal 1225 26,600 bps with Date 1967 to 1984

Product lines like Oyster Precision (no date), Oysterdate, Date, some early Speedkings and Air Kings contain these 1200-series manual movements. All had waterproof Oyster cases offered in three sizes. In sales catalogs these were described as Ladies: 23mm x 8.5mm, Mid-Size: 30 mm x 11 mm, and Men's: 34.5mm x 10.3mm

PROS AND CONS

Current prices for these references are soft and there are bargains to be had. The price-to-value proposition is compelling.

These are hand-wound watches and require daily winding. While some owners love this, others don't. If you can spare the minute, it takes each day to turn the crown a dozen times, these pieces will keep you on schedule.

The cases are slimmer and lower-profile than the automatic Perpetual models of similar age. Many claim they're more comfortable to wear and sit lower on the wrist (helpful if you have a pronounced wrist bone). They're equally as water resistant as automatic models and finished to similarly high standards.

There is an urban myth circulating on watch forums suggesting that daily use of the crown causes accelerated deterioration of the crown tube, resulting in crossed threads and a loss of water resistance. Whether or not there's any evidence to support this, crown tubes are inexpensive and easily replaced.

With accuracy and reliability being a non-issue, the joy of owning one of these Oyster Precisions comes down to case size and dial style.

Precision references most commonly feature thin, polished bezels, but can be found with the machine-turned bezels. These thin bezels accentuate their dials, giving the illusion of a larger watch, hence their reputation for wearing larger than their size.

Given their long production run, dials from these watches are available in a wide variety of styles and colors. These range from the arrowhead hour markers and alpha-style hands of the 1950s, to the double-bars and baton hands of the 1980s. Some dials feature unusual and collectible markings, such as the Meters First water resistance rating.

Earlier references have a distinctive vintage look and feel that is subdued, yet iconic. Dials, in particular, reflect the style trends of their era.

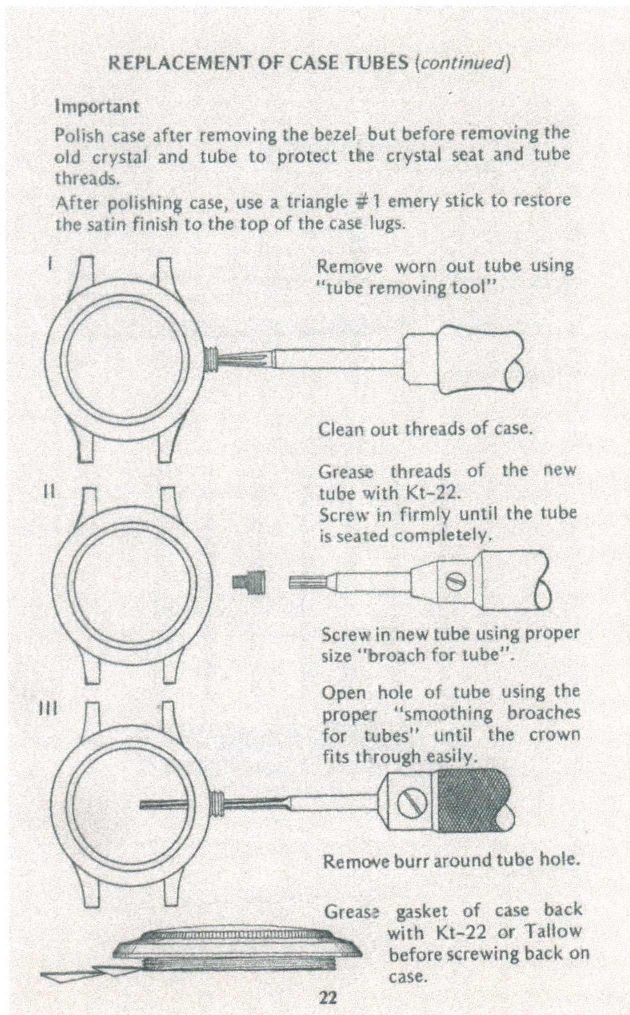

REPLACEMENT OF CASE TUBES *(continued)*

Important

Polish case after removing the bezel but before removing the old crystal and tube to protect the crystal seat and tube threads.

After polishing case, use a triangle #1 emery stick to restore the satin finish to the top of the case lugs.

I

Remove worn out tube using "tube removing tool"

Clean out threads of case.

II

Grease threads of the new tube with Kt-22.

Screw in firmly until the tube is seated completely.

Screw in new tube using proper size "broach for tube".

III

Open hole of tube using the proper "smoothing broaches for tubes" until the crown fits through easily.

Remove burr around tube hole.

Grease gasket of case back with Kt-22 or Tallow before screwing back on case.

22

While significant price appreciation is unlikely, prices have been stable for several years. These watches are not investments, but reasonably-priced, functional jewelry, that happens to be comfortable to wear and fun to collect.

If you like a low-key, classy, vintage Rolex, there are few better references to start with than one of these.

OYSTER PRECISION REFERENCES

Name	Ref	Cal	Bezel	Size	Case	Description
Oyster Precision	**6422**	1215	Polished (Smooth)	Full Size	SS	Round, also "Royal Precision" & "1/4 Century Club"
Oyster Precision	**9083**	1210	Polished (Smooth)	Full Size	SS	Round, "UFO", Integrated bracelet
Oyster Precision	**9665**	Precision	Polished (Smooth)	Full Size	YG	Round
Precision	**2611**	Precision	Polished (Smooth)	Ladies	SS	Square
Precision	**3604**	1210		Full Size	YG	Cushion
Precision	**3667**	1210		Full Size	YG	Round
Precision	**3675**	1210	Polished (Smooth)	Full Size	SS	Round, Sub-seconds (1930s)
Precision	**4309**	1601		Ladies	WG	Square
Precision	**4357**	Precision		Full Size	RGSS	Round
Precision	**4410**	700		Full Size	YG	Round, Sub-seconds
Precision	**4470**	700		Full Size	YG	Square, Sub-seconds
Precision	**4516**	Precision		Full Size	YG	Round, Fancy lugs
Precision	**6424**	1215	Polished (Smooth)	Full Size	SS	Precision
Precision	**6430**	1225	Polished (Smooth)	Mid Size	SS	Precision
Precision	**8940**	Precison		Full Size	YG	Round
Precision	**8952**	Precision	Engine Turned	Full Size	YG	Round
Precision	**9002**	Precision		Full Size	SS	Round
Precision	**9022**	1210		Full Size	SS	Round
Precision	**9081**	Chronometer		Full Size	YG	Round, Sub-seconds
Precision	**9106**	1210		Full Size	YG	Round, "UFO", Integrated bracelet
Precision	**9291**	Precision		Ladies	YG	Round
Precision	**9659**	1210		Full Size	YG	Round
Precision	**9708**	1220	Polished (Smooth)	Full Size	YG	Round
Precision	**9829**	Precision		Full Size	SS	Round
Precision	**9996**	Precision		Full Size	YG	Round

SS : Stainless Steel | YG : Yellow Gold | RG: Rose Gold WG | White Gold | LN : Lunette Noir

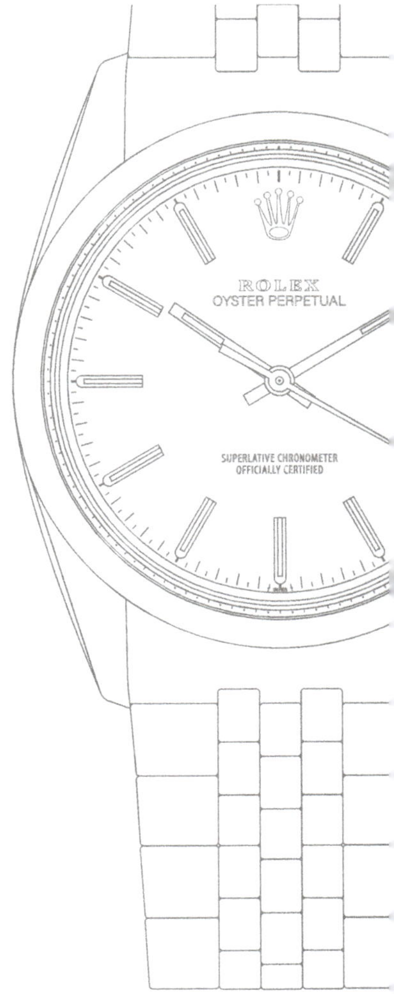

OYSTER PERPETUAL

The story of the Oyster Perpetual is intimately related to that of the Bubbleback from which it evolved. From the 1950s onward, the Rolex Oyster Perpetual collection has been positioned and priced as the entry-level Rolex. The simple three-hand design, devoid of embellishments or complications, continues to be a commercially successful formula.

"Perpetual" is a trademark reserved and used for the automatic self-winding movement. This innovation emerged in the 1930s and was not exclusive to Rolex. Other Swiss watchmakers were designing self-winding movements with sleeker and more integrated solutions, allowing slim case profiles.

The Rolex approach was to retrofit an auto-winding rotor to existing manual winding movements. This approach allowed for a manual wind to top up the reserved power, and provided owners with a level of reassurance and comfort that other fully automatic watches did not. The legacy manual-wind capability was skillfully exploited and marketed as a desirable feature. While not the most technically sophisticated solution, it proved a winner commercially.

It also had the added advantage of being entirely familiar to watchmakers accustomed to the manual-winding movement that lay beneath the rotor. The Perpetual proved not only robust but simple and inexpensive to service.

Dial designs have evolved subtly, reflecting tastes and styles across the decades. Collectors today have a vast range to select from, depending on the decade of interest. Most popular are the textured waffle dials with the Explorer-style 3-6-9 hour markers. These reflect the design aesthetic of the 1950s and include arrowhead or shark tooth-shaped hour markers.

Examples from this period usually have dauphine or alpha-shaped hands.

When choosing a watch from this collection, be sure the style and color of the hands match the dial furniture. With so many in circulation, many examples have been customized to the particular tastes of their owners.

There are approximately 140 references in the Oyster Perpetual line, offered in Full-Size (38mm), Mid-Size (32-34mm) and Ladies-Size (26-32mm). There are a further 80 references across the Oyster Perpetual

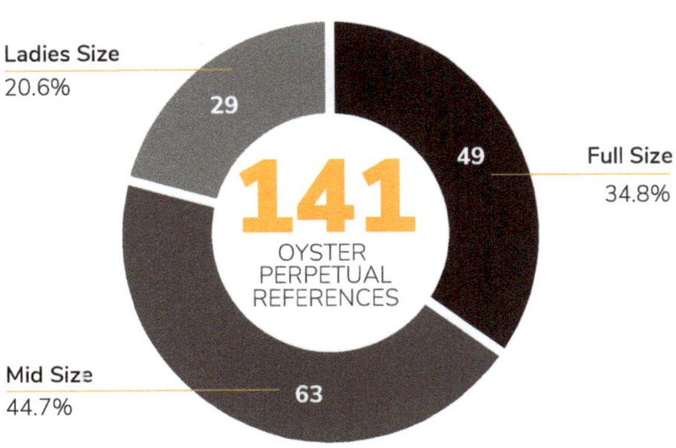

OYSTER PERPETUAL MOVEMENTS

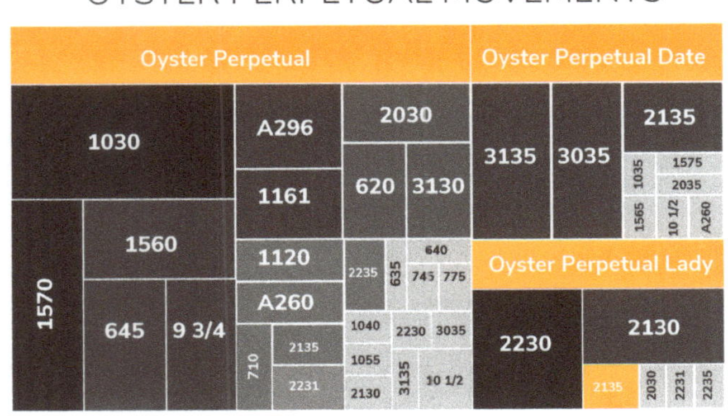

Date, Oyster Perpetual Lady, and Oysterdate models.

OYSTER PERPETUAL REFERENCES

Ref	Cal	Bezel	Size	Description
1003	1570	Polished (Smooth)	Full Size	
1004	1570	Engine Turned	Full Size	
1005	1570	Fluted	Full Size	
1006	1570	Polished (Smooth)	Full Size	
1007	1560	Engine Turned	Full Size	
1008	1560	Engine Turned	Full Size	
1009	1570	Engine Turned	Full Size	
1010	1560	Polished (Smooth)	Full Size	
1011	1570	Fluted	Full Size	
1012	1560	Polished (Smooth)	Full Size	
1013	1560	Fluted	Full Size	
1014	1560	Polished (Smooth)	Full Size	
1018	1560	Fluted	Full Size	
1020	1560	Polished (Smooth)	Full Size	
1024	1560	Polished (Smooth)	Full Size	
1025	1570	Polished (Smooth)	Full Size	
1035			Full Size	
1420			Full Size	
1423			Full Size	
3064	620	Polished (Smooth)	Full Size	
3458	9¾"	Engraved	Full Size	
3496			Full Size	
3794	9¾"		Full Size	
4021			Full Size	
4984			Full Size	
5028	A296	Polished (Smooth)	Full Size	
5052			Full Size	
5065			Full Size	
6021			Full Size	
6023	105"	Engine Turned	Full Size	
6088	A296	Polished (Smooth)	Full Size	
6098	775	Polished (Smooth)	Full Size	
6099	A296	Engine Turned	Full Size	
6110	A260	Polished (Smooth)	Full Size	
6119			Full Size	
6202	A296	Numbered	Full Size	
6210	10½	Polished (Smooth)	Full Size	
6290	640	Polished (Smooth)	Full Size	
6301	1570	Fluted	Full Size	
6332	A296	Polished (Smooth)	Full Size	
6598	745	Polished (Smooth)	Full Size	
6599	1030	Fluted	Full Size	
6611	1055	Fluted	Full Size	
6614	1030	Polished (Smooth)	Full Size	
6634	1030	Polished (Smooth)	Full Size	
14203	3035	Fluted	Full Size	
14233	3130	Fluted	Full Size	
14238	3130	Fluted	Full Size	
116000	3130	Polished (Smooth)	Full Size	
116619LB	3135	Numbered	Full Size	
3272			Ladies	
3352			Ladies	
3686			Ladies	
4686			Ladies	
6504	1120	Polished (Smooth)	Ladies	
6507	1120	Polished (Smooth)	Ladies	
6509	1120	Fluted	Ladies	
6618	1161	Polished (Smooth)	Ladies	
6619	1161	Fluted	Ladies	
6623	1161	Engine Turned	Ladies	
6718	2030	Polished (Smooth)	Ladies	
6719	2130	Fluted	Ladies	
6723	2030	Polished (Smooth)	Ladies	
6724	2030	Engine Turned	Ladies	
7603			Ladies	
7608			Ladies	
7609			Ladies	
7618			Ladies	
7619			Ladies	
7623			Ladies	
7624			Ladies	
76243	2135	Engine Turned	Ladies	
79160	2230	Engine Turned	Ladies	

Ref	Cal	Bezel	Size	Description
79190	2235	Engine Turned	Ladies	
79240	2235	Engine Turned	Ladies	
176234	2231	Fluted	Ladies	
177200	2231	Polished (Smooth)	Ladies	
177210	2135	Engine Turned	Ladies	
761833			Ladies	
1023		Fluted	Mid Size	
1038	1570	Engine Turned	Mid Size	
1507	1570	Engine Turned	Mid Size	
3716	635	Polished (Smooth)	Mid Size	
4270	710	Polished (Smooth)	Mid Size	
4362	9¾"	Polished (Smooth)	Mid Size	
4857			Mid Size	
5016	620	Polished (Smooth)	Mid Size	
5018	9¾"	Polished (Smooth)	Mid Size	
5173			Mid Size	
5552	1030	Polished (Smooth)	Mid Size	
6011	9¾"	Engine Turned	Mid Size	
6016			Mid Size	
6018	9¾"	Polished (Smooth)	Mid Size	
6050	620	Polished (Smooth)	Mid Size	
6082	710	Polished (Smooth)	Mid Size	
6084	620	Polished (Smooth)	Mid Size	
6085	645	Engine Turned	Mid Size	
6090	A260	Polished (Smooth)	Mid Size	
6092	A260	Polished (Smooth)	Mid Size	
6102	645	Polished (Smooth)	Mid Size	
6107	645	Engine Turned	Mid Size	
6284	645	Engine Turned	Mid Size	
6285	645	Engine Turned	Mid Size	
6303	645	Engine Turned	Mid Size	
6334	645	Polished (Smooth)	Mid Size	
6342			Mid Size	
6532	1030	Polished (Smooth)	Mid Size	
6539	A296	Mid Size	Mid Size	
6546	1030	Polished (Smooth)	Mid Size	

Ref	Cal	Bezel	Size	Description
6549	1161	Polished (Smooth)	Mid Size	
6551	1161	Fluted	Mid Size	
6556	1040	Polished (Smooth)	Mid Size	
6564	1030	Polished (Smooth)	Mid Size	
6565	1030	Engine Turned	Mid Size	
6566	1030	Polished (Smooth)	Mid Size	
6567	1030	Fluted	Mid Size	
6569	1030	Engine Turned	Mid Size	
6580	1030	Polished (Smooth)	Mid Size	
6582	1030	Engraved	Mid Size	
6584	1030	Polished (Smooth)	Mid Size	
6585	1030	Polished (Smooth)	Mid Size	
6590	1030	Polished (Smooth)	Mid Size	
6593	1030	Fluted	Mid Size	
6748	2030	Polished (Smooth)	Mid Size	
6751	2030	Fluted	Mid Size	
7701			Mid Size	
7708			Mid Size	
7748			Mid Size	
7751			Mid Size	
8011		Polished (Smooth)	Mid Size	Air Giant
8053		Engine Turned	Mid Size	Canadian Market
8055		Engine Turned	Mid Size	Canadian Market
8058		Engine Turned	Mid Size	Canadian Market
8074		Polished (Smooth)	Mid Size	Canadian Market
8075		Engine Turned	Mid Size	Canadian Market
8076		Polished (Smooth)	Mid Size	Canadian Market
8077		Engine Turned	Mid Size	Canadian Market
8078		Polished (Smooth)	Mid Size	Canadian Market
8079		Engine Turned	Mid Size	Canadian Market
8080		Polished (Smooth)	Mid Size	Canadian Market
14208	3130	Polished (Smooth)	Mid Size	
115234	3135	Fluted	Mid Size	
14203M	3130	Polished (Smooth)	Mid Size	
14208M	3130	Polished (Smooth)	Mid Size	

OYSTER PERPETUAL DATE

Sometimes known as the Rolex Date it is often confused with the larger Datejust. It made its debut in the 1950s following the 1945 launch of the larger Datejust. At 34mm, it is 2mm smaller than the Datejust but the same size as the Oysterdate and Air-King from which it was born.

The smaller case allows for a narrower 19mm lug width and subsequently a thinner bracelet. The Rolex Oyster Perpetual Date is available on both Oyster and Jubilee bracelets and has a wide variety of bezel and dial combinations.

The modern version is known as the Rolex Oyster Perpetual Date 34 and belongs to the Datejust Collection. Early predecessors like the Oysterdate were considered independent product lines and not part of the Datejust collection.

OYSTER PERPETUAL DATE REFERENCES

Ref	Cal	Bezel	Size	Case	Description
1503	1570	Fluted	Mid Size	YG	Round
1560	1565	Polished (Smooth)	Mid Size		Round
1565	1575	Polished (Smooth)	Mid Size		Round
4467	10½	Polished (Smooth)	Full Size		Round
5075	1035	Engine Turned	Mid Size	YGSS	Round
6335	A260		Mid Size		Round
6534	1035	Polished (Smooth)	Mid Size	SS	Round, also YG
6917	2035	Fluted	Ladies		Round
15000	3035	Polished (Smooth)	Mid Size	SS	Round
15010	3035	Engine Turned	Mid Size	SS	Round
15037	3035	Fluted	Mid Size	YG	Round
15038	3035	Fluted	Mid Size	YG	Round
15053	3035	Engine Turned	Mid Size	YGSS	Round
15058	3035	Engine Turned	Mid Size	YG	Round
15200	3135	Polished (Smooth)	Mid Size	SS	Round
15203	3135	Polished (Smooth)	Mid Size	YGSS	Round
15210	3135	Engine Turned	Mid Size	SS	Round
15223	3135	Fluted	Mid Size	YGSS	Round
15233	3135	Engine Turned	Mid Size	YGSS	Round
15238	3135	Fluted	Mid Size	YG	Round
15505	3035	Polished (Smooth)	Mid Size	GS	Round
69160	2135	Polished (Smooth)	Ladies	SS	Round
69174	2135	Fluted	Ladies	SS	Round
69190	2135	Engine Turned	Ladies	SS	Round
69240	2135	Engine Turned	Ladies	SS	Round
115200	3135	Polished (Smooth)	Mid Size	SS	Round
115210	3135	Engine Turned	Mid Size	SS	Round
115234	3135	Fluted	Mid Size	SS	Round

SS : Stainless Steel | YG : Yellow Gold

OYSTER PERPETUAL LADY REFERENCES

Ref	Cal	Bezel
54439	2030	Fluted
67180	2130	Polished (Smooth)
67183	2130	Polished (Smooth)
67188	2130	Polished (Smooth)
67190	2130	Jewel Set
67193	2130	Fluted
67194	2130	Fluted
67197	2130	Fluted
67198	2130	Fluted
67230	2135	Engine Turned
67233	2130	Engine Turned
69628	2135	Numbered
76030	2230	Engine Turned
76080	2230	Polished (Smooth)
76094	2230	Fluted
76180	2230	Fluted
76183	2230	Polished (Smooth)
76188	2230	Polished (Smooth)
76193	2230	Fluted
76198	2230	Fluted
76233	2230	Engine Turned
77438	2230	Fluted
169628	2235	Numbered
176200	2230	Polished (Smooth)
176210	2231	Engine Turned

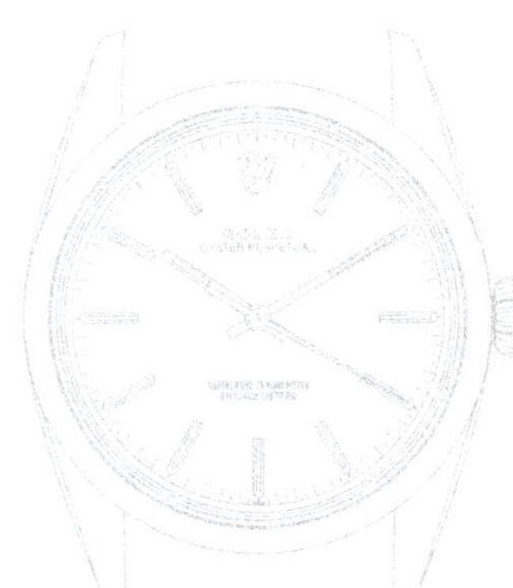

OYSTERDATE

The Oysterdate name was applied to 34mm midsize Oyster cases starting as early as 1951. By 1966, it became "Oyster Perpetual – Date" before losing the hyphens in the late 1970s and becoming "Oyster Perpetual Date". Use of the Oysterdate name seems to have entirely disappeared by 1970.

The Oysterdate name was applied liberally to several classes of watches. Most were manual-winding Precision-class movements like the ref. 6694. A few had automatic chronometer-grade, Perpetual movements, with "Officially Certified Chronometer" or "Ritetime" on the dial (e.g., ref. 6518). The only trait they had in common were their 32mm-34mm cases, which came to be known as Boys Size and Mid Size. The Oysterdate is the original Mid Size Rolex.

The date complication is slow-setting, meaning that if the watch goes unwound and stops, then setting the correct date involves winding the crown the long way around past midnight.

The Oysterdate date wheel, with its open 6s and 9s, is known as "roulette style", and has the odd dates printed in black and even dates printed in red.

The reason behind the roulette design is a mystery. Urban legend suggests it originates from post-war rationing when certain supplies were available on alternate days, and this convenient arrangement helped track them properly. Whatever the reason, it is an attractive and unusual feature.

Today, collectors seek out the Oysterdate for their rarity as well as the 1950s design aesthetic on their dials. These details include alpha and dauphine hands and tapered indexes (arrowhead, shark tooth and diamond-shaped). Waffle textured dials and Explorer 3-6-9 indexes are also correct and are as popular today as they were in the 50s.

OYSTERDATE PERPETUAL

The Oysterdate Perpetual such as the ref. 6518 uses cal. 1035 with its beautiful butterfly-style rotor. Caliber 1035 is a chronometer-grade movement and timed to COSC standards, and their dials will bear the "Officially Certified Chronometer" text. This caliber is the same 1035 used in the early GMT-Master and Submariner, and both of these Professional watches command considerably higher prices. Unfortunately, Oysterdate Perpetuals are being cannibalized to provide parts for these other, more expensive models. They're already rare and becoming more so.

OYSTERDATE REFERENCES

Name	Ref	Cal	Bezel	Size	Case
Oysterdate	**1500**	1570	Polished (Smooth)	Mid Size	SS
Oysterdate	**1501**	1570	Engine Turned	Mid Size	SS
Oysterdate	**1505**	1570	Fluted	Mid Size	
Oysterdate	**1507**	1570	Other	Mid Size	
Oysterdate	**1550**	1570	Polished (Smooth)	Mid Size	
Oysterdate	**1570**	1570	Fluted	Full Size	
Oysterdate	**1600**	1570	Polished (Smooth)	Full Size	
Oysterdate	**1601**	1570	Fluted	Full Size	
Oysterdate	**1603**	1570	Fluted	Full Size	
Oysterdate	**6075**	295	Engine Turned	Full Size	
Oysterdate Precision	**6094**		Smooth	Full Size	SS
Oysterdate	**6423**	1210	Engine Turned	Full Size	
Oysterdate	**6466**	1210	Polished (Smooth)	Mid Size	
Oysterdate	**6494**	1210	Polished (Smooth)	Mid Size	SS
Oysterdate Perpetual	**6518**	1035	Polished (Smooth)	Mid Size	SS
Oysterdate	**6519**	1135	Polished (Smooth)	Ladies	SS
Oysterdate	**6522**	1100	Polished (Smooth)	Ladies	
Oysterdate	**6523**	1100	Engine Turned	Ladies	
Oysterdate	**6524**	1210	Polished (Smooth)	Mid Size	SS
Oysterdate	**6525**	1100	Fluted	Ladies	SS
Oysterdate	**6534**	1030	Polished (Smooth)	Mid Size	SS
Oysterdate	**6605**	1065	Fluted	Full Size	YG
Oysterdate	**6627**	1161	Fluted	Mid Size	YG
Oysterdate	**6694**	1225	Polished (Smooth)	Mid Size	SS
Oysterdate	**6964**	1210	Polished (Smooth)	Mid Size	SS

SS : Stainless Steel | YG : Yellow Gold

OYSTERQUARTZ

Despite this being a quirky and unusual reference, it played a significant role in Rolex history. Rolex had been researching electronic watches since the early 1950s and was awarded their first electro-mechanical watch patent in 1952.

Rolex was awarded 50 patents from 1960 to 1990. 21 of these patents were for electronic watches and digital displays. Rolex's investment in quartz technology was significant.

Rolex introduced its first commercially available quartz watch in 1970. The Quartz Date 5100 shared the Beta 21 movement used by other Swiss companies, including Omega and Enicar.

Assembly of the Beta 21 took place across multiple manufacturing facilities, with limited modifications for each brand's specifications. Sixteen different Swiss watch companies began selling Beta 21 quartz watches in 1970, including Rolex with its Quartz Date 5100. After making only 1,000 units of the Beta 21, Rolex started to develop their quartz movement for the watch that would become the Oysterquartz.

In 1977, Rolex introduced their first entirely in-house quartz movements for the Datejust (ref. 5035) and Day-Date (ref. 5055) Oysterquartz models. These were COSC certified, which is very unusual for quartz movements.

At launch, these were remarkable technological achievements created with a level of fit and finish befitting the milestone achievement.

Rolex designed and built these 11 jewel movements like mechanical movements with bridges, plates, and lever escapements. They used the latest CMOS circuitry, a 32khz oscillator, and an analog thermal compensator.

Quartz movements are vulnerable to extreme temperatures, and a trimmer feature was incorporated to allow watchmakers to compensate for drift of the quartz crystals. It is very unusual to

find quartz movements designed to be watchmaker serviceable in this way.

Rolex offered many dial options. These include malachite, lapis, silver, blue, black, champagne, gold, tapestry stripes, diamond-set, wood, and others, including Stella enamel in several colors,. Most of these exotic dials were for the premium Day-Date Oysterquartz, which was also available with jeweled dials, bezels, and bracelets.,

The Oysterquartz models ran for 25 years, and Rolex made fewer than 25,000 units. Rolex dropped the ref. 17000 from their catalog in 2002 but continued to offer the two-tone and solid gold models until 2003, when stocks were finally depleted.

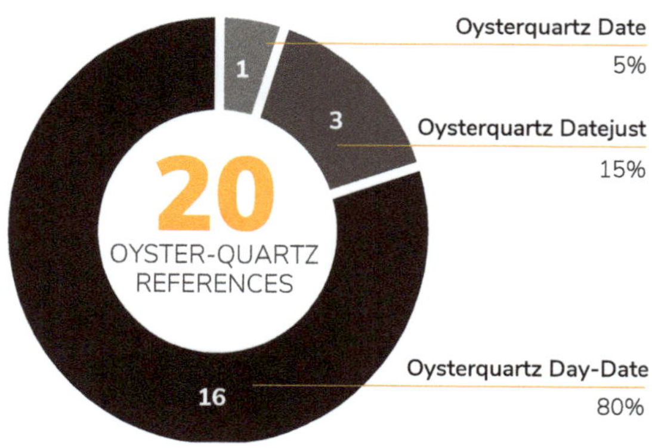

THE FIRST QUARTZ DATE REF. 5100

The ref. 5100 emerged from a co-development and collaboration agreement called Centre Electronique Horloger (CEH). The resulting quartz movement called "Beta 21" was deployed straight into the Quartz Date ref. 5100.

It featured an integrated bracelet and a sapphire crystal, also firsts for Rolex. Rolex could not fit the Beta 21 movement into the Oyster case without modifications. As such, it was described as water resistant and not waterproof. The Rolex Quartz Date never received a depth rating.

The 5100 debuted on June 5, 1970. Initial orders exceeded expectation, with the planned production run of 1,000 watches selling out before production even began. All 1,000 serially-numbered 5100s were sold between 1970 and 1972. Today, collectors highly prize a ref.5100 complete with box and papers.

In 1972 Rolex withdrew from CEH and the Beta 21 consortium. They began developing a quartz movement of their design and the watch that would house it, the Oysterquartz.

OYSTERQUARTZ REFERENCES

Name	Ref	Cal	Start	End	Bezel	Case	Description
Oysterquartz Date	**5100**	Beta 21	1970	1972	Fluted	YG	Very first Oysterquartz model
Oysterquartz Datejust	**17000**	5035	1977	2001	Polished (Smooth)	SS	Very last of the Oysterquartz models Non-COSC
Oysterquartz Datejust	**17013**	5035	1977	2001	Fluted	SSYG	Jubilee bracelet Non-COSC
Oysterquartz Datejust	**17014**	5035	1977	2001	Fluted	WG	Jubilee bracelet
Oysterquartz Day-Date	**1901**	5055			Fluted		
Oysterquartz Day-Date	**1902**	5055			Pyramid		
Oysterquartz Day-Date	**1903**	5055			Pyramid		Diamond
Oysterquartz Day-Date	**1904**	5055					Diamond
Oysterquartz Day-Date	**1905**	5055					Baguette diamond
Oysterquartz Day-Date	**1907**	5055					
Oysterquartz Day-Date	**1914**	5055					
Oysterquartz Day-Date	**1916**	5055					
Oysterquartz Day-Date	**19018**	5055	1977	2001	Fluted	YG	Gem set dials, bezels and bracelets
Oysterquartz Day-Date	**19019**	5055	1977	2001	Fluted	WG	Gem set dials, bezels and bracelets
Oysterquartz Day-Date	**19028**	5055	1977	2001	Pyramid	YG	Gem set dials, bezels and bracelets Pyramid hour markers
Oysterquartz Day-Date	**19038**	5055	1977	2001	Pyramid	YG	Gem set dials, bezels and bracelets Pyramid hour markers and pyramid bracelet
Oysterquartz Day-Date	**19048**	5055	1977	2001		YG	Gem set dials, bezels and President bracelets
Oysterquartz Day-Date	**19049**	5055	1977	2001		WG	Gem set dials, bezels and President bracelets
Oysterquartz Day-Date	**19068**	5055	1977	2001		YG	Gem set dials, bezels and President bracelets
Oysterquartz Day-Date	**19148**	5055	1977	2001		WG	Gem set dials, bezels and Karat bracelets

SS : Stainless Steel | YG : Yellow Gold | RG : Rose Gold | WG : White Gold

OYSTER PERPETUAL DATEJUST

The first Datejust was ref. 4467, introduced in 1945. While it officially marked the company's 40th anniversary, it also unofficially marked the end of WWII. It is remarkable that Rolex was able to design, develop, and manufacture this watch amid the chaos of the war that raged throughout Europe.

The name Datejust is a reference to the quick-changing date, which jumped to the next date on the stroke of midnight, rather than rolling over slowly. As such, the date was always just right. This seemingly simple innovation was a milestone development in the watch industry at the time.

The ref. 4467 also introduced the fluted-style bezel and the new Jubilee bracelet. The iconic cyclops magnifier over the date aperture was added nine years later in 1954.

The 36mm size was substantial for the time, and Rolex later added a Mid-Size 34mm version and a Ladies size 28mm. These early examples were Bubblebacks with convex casebacks. It wasn't until the introduction of cal. 1065 in 1957 that the Datejust got its flat caseback.

The Datejust was a commercial success, thanks to the accurate COSC certified Perpetual movement, waterproof Oyster case and the new-fangled date complication. Rolex went on to make approximately 280 variations of the Datejust in three case sizes, using 25 different movements and numerous bezel combinations.

An Oyster Datejust Bubbleback sat on the wrist of Chuck Yeager when he broke the sound barrier in 1947 flying the Bell X-1 rocket plane.

In 2009 Rolex released the Oyster Perpetual Datejust II in a larger 41mm Oyster case. In 2016, this was followed by the Oyster Perpetual Datejust 41. While the Oyster case size was the same, it had smaller hour indexes and a thinner bezel, giving the illusion of a much larger watch.

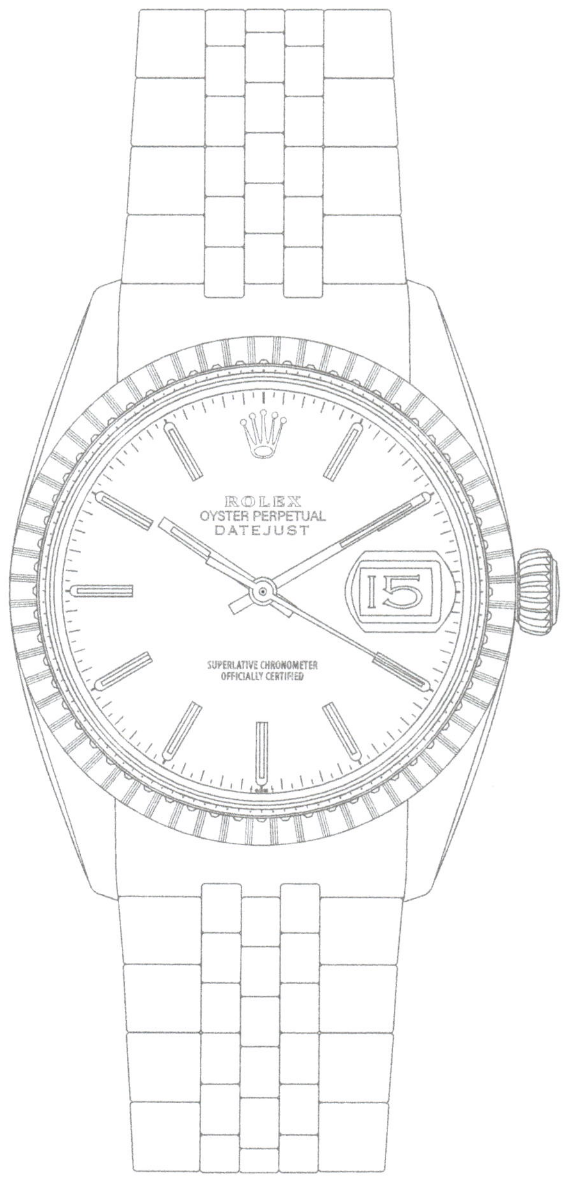

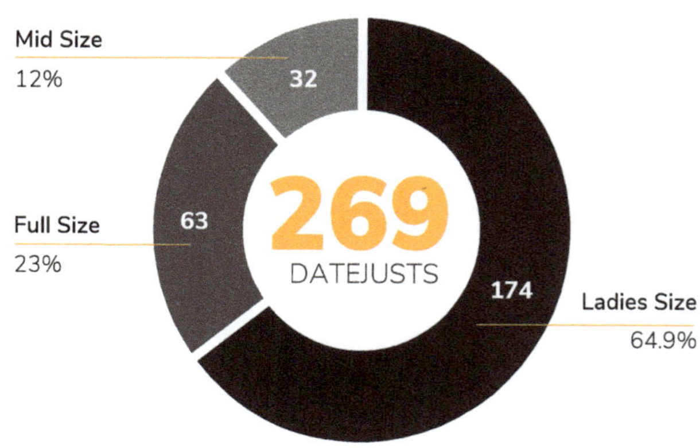

To say the Datejust model line is long-lived is an understatement. Rolex made these models in large quantities, and the pre-owned market is awash with them. They have numerous dial variations reflecting the design aesthetic of the era. A few of these dials can be found with the prized red text and in some cases waterproof depth ratings too. Dial textures include waffle, striped, tapestry, linen, and embossed. The pie-pan dial is a popular variant with vintage collectors, as are the rare all-stainless steel versions. The ref. 6105 is particularly unusual, as it was produced for left-handed individuals, with the winding crown at the 9 o'clock position.

When buying a vintage Datejust, you'd be well-advised not to obsess over originality and correctness, as these watches were often customized by the AD when first sold. Dealers would swap dials and hands at a customer's request. Look for one in good condition and running order, then select what appeals to you aesthetically. If you follow the advice in this guide, these can be well-bought at modest prices.

*Prototype Bell X-1 rocket plane, undergoing shakedown tests.
An early Oyster Datejust Bubbleback sat on the wrist of Chuck Yeager
when he broke the sound barrier in 1947*

Blue dials are particularly popular and this example ref 16233 is typical of what is considered, good condition. Opinions are divided about the two tone aesthetic of the Datajust. In Asia it's nicknamed the Uncle Watch for its garish 1980's vibe. Among western collectors it's considerd either timeless and classic, or ostentatious and crass. Whatever your taste, there is bound to be a configuration to suit you. It is a robust and versatile watch, bursting with iconic Rolex design cues..

ROLEX OYSTER DATEJUST REF. 1601 "THE WIDE BOY"

These two contrasting examples illustrate condition from Good to Typical. The stainless steel example has a sharp and crisp case profile with matching hands showing some patina.

The two-tone example has a poorly refinished dial, after-market hands, and a heavily polished case. The hour marker tritium has been removed and appears in stark contrast to the hands. Bottom right shows final professional restoration.

DATEJUST REFERENCES

Ref	Cal	Bezel	Size
1600	1570	Fluted	Full Size
1601	1570	Fluted	Full Size
1603	1570	Engine Turned	Full Size
1605	1570	Engine Turned	Full Size
1607	1570	Bark	Full Size
1611	1570	Bark	Full Size
1620	1570	Polished (Smooth)	Full Size
1622	1570	Engine Turned	Full Size
1623	1570	Fluted	Full Size
1624	1570	Fluted	Full Size
1630	1570	Fluted	Full Size
5030	710	Polished (Smooth)	Full Size
5031	710	Engine Turned	Full Size
6031	710	Fluted	Full Size
6074	710	Polished (Smooth)	Full Size
6075	A295	Fluted	Full Size
6104	A296	Polished (Smooth)	Full Size
6105	745	Engine Turned	Full Size
6155	1570	Engine Turned	Full Size
6304	A296	Polished (Smooth)	Full Size
6305	A296	Fluted	Full Size
6604	1065	Polished (Smooth)	Full Size
6605	1065	Fluted	Full Size
6909	2030	Engine Turned	Full Size
6914	2135	Engine Turned	Full Size
6923	2030	Engine Turned	Full Size
7518	234	Engine Turned	Full Size
8029	390	Polished (Smooth)	Full Size
16000	3035	Fluted	Full Size
16013	3035	Fluted	Full Size
16014	3035	Engine Turned	Full Size
16018	3035	Fluted	Full Size
16030	3035	Fluted	Full Size
16078	3035	Bark	Full Size
16200	3135	Polished (Smooth)	Full Size
16203	3135	Polished (Smooth)	Full Size

Ref	Cal	Bezel	Size
16220	3135	Engine Turned	Full Size
16233	3135	Fluted	Full Size
16234	3135	Fluted	Full Size
16238	3135	Fluted	Full Size
18958	5055		Full Size
68188	2135	Fluted	Full Size
79079	2235	Polished (Smooth)	Full Size
116034	3130	Fluted	Full Size
116138	3135	Fluted	Full Size
116139	3135	Fluted	Full Size
116188	3135	Jewel Set	Full Size
116189	3135	Jewel Set	Full Size
116199	3135	Jewel Set	Full Size
116200	3135	Polished (Smooth)	Full Size
116201	3135	Polished (Smooth)	Full Size
116203	3135	Polished (Smooth)	Full Size
116208	3135	Polished (Smooth)	Full Size
116231	3135	Fluted	Full Size
116233	3135	Fluted	Full Size
116234	3135	Fluted	Full Size
116238	3135	Fluted	Full Size
116243	3135	Jewel Set	Full Size
118399	3235	Jewel Set	Full Size
126301	3235	Polished (Smooth)	Full Size
126303	3235	Polished (Smooth)	Full Size
126331	3235	Fluted	Full Size
126333	3235	Fluted	Full Size
6251	710	Fluted	Ladies
6527	1160	Bark	Ladies
6824	2030	Polished (Smooth)	Ladies
6825	2035	Engine Turned	Ladies
6826	2030	Fluted	Ladies
6828	2035	Jewel Set	Ladies
6900	2030	Fluted	Ladies
6901	2035	Bark	Ladies
6902	2030	Bark	Ladies

Ref	Cal	Bezel	Size
6903	2135	Fluted	Ladies
6906	2135	Jewel Set	Ladies
6907	2030	Jewel Set	Ladies
6908	2035	Fluted	Ladies
6910	2135	Jewel Set	Ladies
6911	2030	Jewel Set	Ladies
6912	2030	Fluted	Ladies
6913	2030	Jewel Set	Ladies
6915	2135	Jewel Set	Ladies
6917	2030	Fluted	Ladies
6925	2030	Jewel Set	Ladies
6926	2135	Jewel Set	Ladies
6927	2030	Jewel Set	Ladies
6928	2135	Jewel Set	Ladies
6930	2030	Fluted	Ladies
6931	2135	Engine Turned	Ladies
6935	2030	Fluted	Ladies
7828	2135	Fluted	Ladies
7906	2030	Fluted	Ladies
7907	710	Fluted	Ladies
7908	2030	Jewel Set	Ladies
7912	1157	Jewel Set	Ladies
7913	1156	Jewel Set	Ladies
7915	2030	Jewel Set	Ladies
7918	2135	Engine Turned	Ladies
7927	1475	Fluted	Ladies
8030	390	Fluted	Ladies
8031	1475	Polished (Smooth)	Ladies
8032	1156	Jewel Set	Ladies
8035	3035	Engine Turned	Ladies
16239	3135	Jewel Set	Ladies
18029	2130	Jewel Set	Ladies
67243	2130	Engine Turned	Ladies
68158	2135	Jewel Set	Ladies
68159	2135	Jewel Set	Ladies
68238	2135	Jewel Set	Ladies

Ref	Cal	Bezel	Size
68240	2135	Polished (Smooth)	Ladies
68243	2135	Polished (Smooth)	Ladies
68246	2135	Polished (Smooth)	Ladies
68258	2135	Jewel Set	Ladies
68266	2135	Jewel Set	Ladies
68268	2135	Jewel Set	Ladies
68273	2135	Fluted	Ladies
68274	2135	Fluted	Ladies
68278	2135	Fluted	Ladies
69068	2135	Jewel Set	Ladies
69069	2135	Jewel Set	Ladies
69078	2135	Jewel Set	Ladies
69079	2135	Jewel Set	Ladies
69088	2135	Jewel Set	Ladies
69089	2135	Jewel Set	Ladies
69126	2135	Jewel Set	Ladies
69128	2135	Jewel Set	Ladies
69136	2135	Jewel Set	Ladies
69138	2135	Jewel Set	Ladies
69139	2135	Jewel Set	Ladies
69158	2135	Jewel Set	Ladies
69160	2135	Polished (Smooth)	Ladies
69163	2135	Polished (Smooth)	Ladies
69166	2135	Polished (Smooth)	Ladies
69168	2135	Jewel Set	Ladies
69173	2135	Fluted	Ladies
69174	2135	Fluted	Ladies
69178	2135	Jewel Set	Ladies
69190	2135	Engine Turned	Ladies
69198	2135	Jewel Set	Ladies
69240	2135	Jewel Set	Ladies
69258	2135	Jewel Set	Ladies
69268	2135	Jewel Set	Ladies
69278	2135	Engine Turned	Ladies
69279	2135	Fluted	Ladies
69288	2135	Jewel Set	Ladies

DATEJUST REFERENCES CONT.

Ref	Cal	Bezel	Size
69298	2135	Jewel Set	Ladies
69299	2135	Jewel Set	Ladies
69308	2135	Jewel Set	Ladies
69318	2135	Jewel Set	Ladies
76243	2230	Fluted	Ladies
78246	2235	Polished (Smooth)	Ladies
78248	2235	Polished (Smooth)	Ladies
78266	2235	Engine Turned	Ladies
78273	2235	Fluted	Ladies
78274	2235	Fluted	Ladies
78278	2235	Jewel Set	Ladies
78279	2235	Jewel Set	Ladies
78288	2235	Jewel Set	Ladies
79068	2235	Jewel Set	Ladies
79078	2235	Jewel Set	Ladies
79088	2235	Jewel Set	Ladies
79089	2235	Jewel Set	Ladies
79126	2135	Jewel Set	Ladies
79136	2135	Jewel Set	Ladies
79138	2235	Jewel Set	Ladies
79158	2235	Jewel Set	Ladies
79160	2235	Polished (Smooth)	Ladies
79163	2235	Polished (Smooth)	Ladies
79166	2135	Jewel Set	Ladies
79168	2135	Polished (Smooth)	Ladies
79173	2235	Fluted	Ladies
79174	2135	Fluted	Ladies
79178	2235	Fluted	Ladies
79190	2235	Polished (Smooth)	Ladies
79193	2235	Fluted	Ladies
79240	2235	Fluted	Ladies
80285	2235	Jewel Set	Ladies
80298	2235	Jewel Set	Ladies
80299	2235	Jewel Set	Ladies
80309	2235	Jewel Set	Ladies
80318	2235	Jewel Set	Ladies

Ref	Cal	Bezel	Size
80319	2235	Jewel Set	Ladies
80328	2235	Polished (Smooth)	Ladies
80329	2235	Jewel Set	Ladies
80359	2235	Jewel Set	Ladies
177234	2235	Fluted	Ladies
178158	2235	Jewel Set	Ladies
178159	2235	Jewel Set	Ladies
178238	2235	Fluted	Ladies
178239	2235	Fluted	Ladies
178240	2235	Polished (Smooth)	Ladies
178241	2235	Polished (Smooth)	Ladies
178243	2235	Polished (Smooth)	Ladies
178245	2235	Polished (Smooth)	Ladies
178246	2235	Polished (Smooth)	Ladies
178248	2235	Polished (Smooth)	Ladies
178269	2235	Polished (Smooth)	Ladies
178271	2235	Fluted	Ladies
178273	2235	Fluted	Ladies
178274	2235	Fluted	Ladies
178275	2235	Fluted	Ladies
178278	2235	Fluted	Ladies
178279	2235	Fluted	Ladies
178286	2235	Jewel Set	Ladies
178288	2235	Jewel Set	Ladies
178313	2235	Jewel Set	Ladies
178343	2235	Jewel Set	Ladies
178344	2235	Jewel Set	Ladies
178383	2235	Jewel Set	Ladies
178384	2235	Jewel Set	Ladies
179136	2235	Jewel Set	Ladies
179138	2235	Jewel Set	Ladies
179158	2235	Jewel Set	Ladies
179159	2235	Jewel Set	Ladies
179160	2235	Polished (Smooth)	Ladies
179161	2235	Polished (Smooth)	Ladies
179163	2235	Polished (Smooth)	Ladies

Ref	Cal	Bezel	Size
179165	2235	Polished (Smooth)	Ladies
179166	2235	Jewel Set	Ladies
179168	2235	Polished (Smooth)	Ladies
179171	2235	Fluted	Ladies
179173	2235	Fluted	Ladies
179174	2235	Fluted	Ladies
179175	2235	Fluted	Ladies
179178	2235	Fluted	Ladies
179179	2235	Fluted	Ladies
179239	2235	Fluted	Ladies
179298	2235	Polished (Smooth)	Ladies
179313	2235	Jewel Set	Ladies
179368	2235	Jewel Set	Ladies
179459	2235	Jewel Set	Ladies
197173		Fluted	Ladies
228348	3255	Jewel Set	Ladies
279135	2236	Jewel Set	Ladies
279160	2236	Polished (Smooth)	Ladies
279171	2236	Fluted	Ladies
279173	2236	Fluted	Ladies
279381	2236	Jewel Set	Ladies
6624	1160	Polished (Smooth)	Mid Size
6800	2030	Polished (Smooth)	Mid Size
6815	2030	Polished (Smooth)	Mid Size
6823	2030	Polished (Smooth)	Mid Size
7815	2030	Jewel Set	Mid Size
7823	234	Fluted	Mid Size
7824	2135	Polished (Smooth)	Mid Size
7825	2030	Jewel Set	Mid Size
7826	2030	Jewel Set	Mid Size
7827	2135	Polished (Smooth)	Mid Size
7926	1475	Polished (Smooth)	Mid Size
8065	3035	Engine Turned	Mid Size
8066	1156	Jewel Set	Mid Size
8067	3035	Polished (Smooth)	Mid Size
15505	3035	Polished (Smooth)	Mid Size

Ref	Cal	Bezel	Size
68248	2135	Polished (Smooth)	Mid Size
68279	2135	Fluted	Mid Size
68286	2135	Jewel Set	Mid Size
68288	2135	Jewel Set	Mid Size
77014	2230	Fluted	Mid Size
77080	2230	Polished (Smooth)	Mid Size
77518	2235	Fluted	Mid Size
78240	2235	Jewel Set	Mid Size
78243	2235	Polished (Smooth)	Mid Size
78286	2235	Jewel Set	Mid Size
81158	2235	Jewel Set	Mid Size
81208	2235	Polished (Smooth)	Mid Size
81298	2235	Jewel Set	Mid Size
81315	2235	Jewel Set	Mid Size
81318	2235	Jewel Set	Mid Size
81319	2235	Jewel Set	Mid Size
81338	2235	Jewel Set	Mid Size

DATEJUST TURN-O-GRAPH

In 1955, Rolex introduced the Datejust Turn-O-Graph model with a novel rotating bezel. It was the genesis for the Rolex Submariner and the Rolex GMT-Master. The Turn-O-Graph became popular with pilots as a navigation tool even before the formal launch of the GMT-Master later in 1955.

The first reference was the 6202, which had a steel Oyster case with a gilt and glossy dial, luminous Mercedes-style hands, luminescent hour markers, and a black bezel, calibrated to 60 minutes.

Collectors and scholars acknowledge the ref. 6202 was the first Rolex tool watch, the first aviator's watch, and the first offered in steel and gold two-tone. This luxury version was initially considered too avant-garde and sold in very modest numbers.

In 1954 several design changes were made with the new ref. 6309. A date complication was added, along with a cyclops date magnifier, and a new engine-turned bezel with markers every five and ten minutes. The movement was a revised cal. 743, replacing the original cal. A260.

The US Air Force informally adopted the Turn-O-Graph, awarding it to pilots returning from the Vietnam and Korean Wars. Later, it was formally adopted and issued to the U.S. Air Force's Thunderbird aerobatic team, hence the nickname, Thunderbird.

The Turn-O-Graph was assigned to the Datejust collection in 1954 in an attempt to make way for the Submariner and GMT-Master. In hindsight, this proved a misguided choice as the Turn-O-Graph's timing bezel on a dressy Datejust was an unusual and inconsistent style combination.

The Turn-O-Graph was revised and reissued in 2000, with cal. 3135, a fluted bezel, distinctive red dial text, and a red seconds hand. This reissued version was a dressier, more luxurious take on the original, but it still failed to find its niche in the public's mind.

After 11 references and 58 years, the Turn-O-Graph was finally retired in 2011. It maintains a second-tier place in Rolex history, and never made a big impression on the public, despite being endorsed by fighter pilots.

DATEJUST TURN-O-GRAPH REFERENCES

Ref	Cal	Bezel	Case
1625	1570	Numbered, Engine Turned	SS YG
1626	1570	Numbered, Engine Turned	
6202	A260	Numbered, Black Insert, Red triangle	SS
6309	710	Numbered, Engine Turned	YG
6609	1066	Numbered, Engine Turned	RG
16253	3035	Numbered, Engine Turned	SSYG
16264	3135	Numbered, Engine Turned	SSGW
16268	3035	Numbered, Engine Turned	YG
116261	3135	Numbered, Fluted	SSRG
116263	3135	Numbered, Fluted	SSYG
116264	3135	Numbered, Fluted	SSWG

SS : Stainless Steel | YG : Yellow Gold | RG : Rose Gold | WG : White Gold

OYSTER PERPETUAL DAY-DATE

The Rolex Day-Date is also known as the Day-Date President, having been worn by several heads of state and industry titans. It is a premium luxury line in precious metals only but available in countless configuration combinations. The Day-Date is one of several references offered with jewel-set bezels. The model comes in a variety of combinations, with some being exceptionally rare. Those described here are the most commonly encountered.

Rolex has offered the Day-Date with some exotic finishes including bark and Morellis. These are rare and highly prized. While the larger Day-Date II and Day-Date 40 have proved popular, the 36mm Day-Date is building a cult following among fashionable millennials. The yellow gold pairs well with jewelry and accessories, and is versatile enough to dress up or down. Stylish young collectors are breathing new life and interest into the vintage 36mm Day-Date.

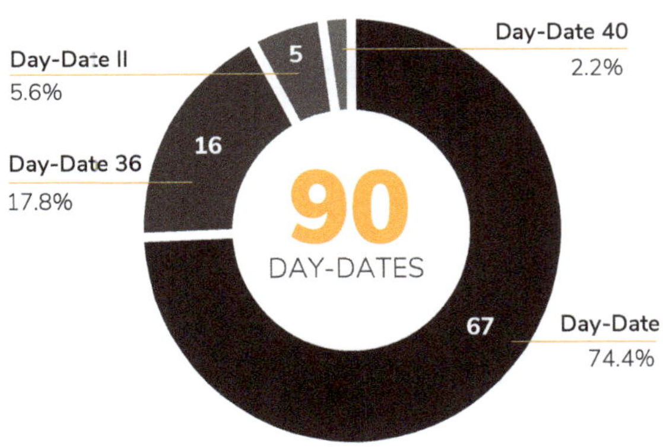

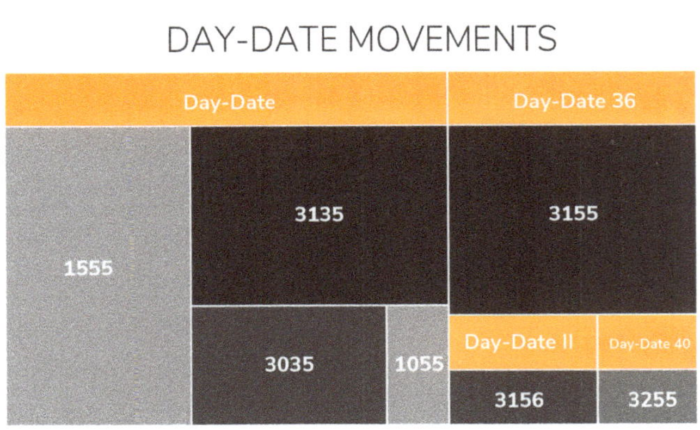

Buying advice

The first two generations (ref. 6511 and ref. 6611 and their variations) are highly sought after collector pieces. If the case condition is good to acceptable, and the dial, hands, and date discs are original, then the only purchase decision is down to price. If the dial is not original or from a later generation walk away.

Early versions of the third generation (ref. 1800) are worth buying. Make sure they have the right dial with the correct markers and matching hands. For later ones, look for unusual dial colors. Precious metals other than yellow gold are unusual and a great buy. Production numbers were considerably higher, so while the first two generations are relatively rare, this generation isn't. With so many in circulation, look for condition and originality.

The Fourth generation (ref. 18000) is not rare, but offered unusual dial materials. Look for Rolex factory dials in stone, wood, and enamel, including rare diamond-set variations.

6500 SERIES (1956 TO 1959)

The Day-Date launched in 1956 with ref. 6510 and ref. 6511. Displaying the day of the week and the date was an industry first. Combined with a precious metal Oyster case, the formula has remained unchanged to this day.

These were only available with white/silver dials, and matching Dauphine hands or Alpha hands. Dial markings include 50m – 165ft, which carried over to the 6511. The Day-Date wording is printed in red and they are typically seen with an Italian day disc.

The 6511 introduced a Day disc in various languages, with English and Italian confirmed. Dial markings can include 50m – 165ft but by the end of the model run, it had been dropped.

The ref. 6511 was replaced by the little known, ref. 6611. This featured an upgraded caliber 1055 earning it the designation "Superlative Chronometer Officially Certified". This replaced the previous text, "Superlative Chronometer by Official Test". It was the first Rolex to feature the text, indicating "especially good results" in the chronometer tests. This model also marked the transition from Dauphine hands to Alpha hands. However, many transition models were made and either is considered correct.

The Rolex Day-Date booklet from the early 1980s

6611 SPECIAL NOTE

For a very brief and indeterminate period, a special Day-Date was made in stainless steel. According to the Antiquorum auction house, the total production ran to only six units, making it the rarest of all Rolex models.

In October 2002, one of these stainless examples fitted with an Oyster bracelet sold for 50,600 CHF. The auction catalog claims it had an engraved caseback with the words "Ecole d'Horlogerie de Genève – 1963", as well as the Rolex logo. These watches were apparently givens as prizes to the best students of the Geneva school of Horology. Rolex never commercialized these stainless steel 6611s, instead, choosing to focus on precious metals.

1800 SERIES (1959 TO 1987)

Rolex introduced an update to the 6500 series in 1959 with the Day-Date ref. 1803. One of the most well-known references, it lasted into the late 1970s. Early 1800 models had cal. 1555. In the 1970s, the 1800 series received a movement upgrade to cal. 1556, which provided a hacking feature.

Early models had alpha-style hands, which transitioned to the batton-style, which was more typical of the 1980's. The Wide-Boy style dial and hands are distinctive, relatively rare, and particularly collectible.

Only the ref. 1804 had a factory-fitted diamond bezel, and it was only available in white gold with 46 brilliant cut diamonds. The ref. 1805 had a diamond and sapphire bezel combination. There are no other identifying marks or hallmarks on the back of either of these bezels.

In the 1960s and 70s, authorized dealers could sell diamond bezels separately to anyone with a Day-Date in the 1800-1811 series. These are genuine factory Rolex parts, but to a collector, a ref. 1804 with a factory-fitted diamond bezel is more desirable than an upgraded 1803.

Combination (multi-colored) and baguette stone bezels were never sold as an upgrade and will only appear on the correct case reference. For example, only ref. 1805 had a diamond and sapphire combination bezel. Only the ref. 1816 had a diamond baguette cut bezel. And only ref. 1817 had an emerald, ruby, sapphire, and diamond combination bezel.

The modern classic Day-Date ref. 18038

If an owner has a problem with a combination jeweled bezel, Rolex will verify the correct case number before working on it. They will not work on an aftermarket or incorrect combination jewel bezel.

In 1977, cal. 3055 arrived in the 18000 series, finally giving the Day-Date quick-setting date capability. This year also saw the launch of the Day-Date Oysterquartz and the cal. 5055 quartz movement. It had a distinctive angular look that set it apart from the mechanical Day-Date.

Bezel options increased on the Crown Collection models, with sapphire, ruby, and emerald in different cuts (like the baguette) and combinations with diamonds. Some had 50 small baguettes and others contained 24 larger baguettes.

Colored stone bezels belong exclusively to the Crown Collection. Rolex strictly controlled these combination bezels, and like the diamond-only bezels, they were not offered for sale by Ads. Colored stone bezels on any other Day-Date reference are aftermarket.

The ref. 18000 men's diamond bezel had 44 brilliant diamonds set in yellow or white gold. The early 18000s had no other distinguishing markings. Later versions had a letter and a number stamped on the back of the bezel, but no Rolex logo or other hallmarks.

1830 SPECIAL NOTE

A special and elusive reference, 1831, was made in no more than eight examples as a custom order in platinum only. It features an OysterQuartz-style case with an integrated bracelet in a King Midas style (neither the Oysterquartz of King Midas was made in platinum). These are very distinctive looking examples with the classic Gérald Genta design aesthetic.

18200 SERIES (1988 TO 2000)

In 1988, Rolex debuted a new Day-Date movement, the cal. 3155, and a new series of Day-Date watches with 18200 reference numbers. This updated movement introduced quickset day capability.

The 18200 series jeweled bezels were platinum, yellow, or white gold, to match the case. It was never offered in rose or pink gold. Rolex began stamping a code on the back of the bezels along with a Rolex hallmark, presumably in response to aftermarket and counterfeit bezels.

The quality and consistency of jewel set bezels improved significantly with this series. They set the standard for classic and modern Day-Dates. The diamonds are always VVS clarity and above E in color. If there are visible inclusions in any of the stones, it is not a Rolex bezel. All stones will be perfectly consistent in color, size, and symmetry, and they will be evenly spaced with the settings perfectly repeated.

There is a rare type of bezel called Cartouche that emerged in the late 1980s. Cartouches had 40 diamonds on Men's Crown Collection and 36 on Ladies'. They have the name Rolex engraved on a plate at the six o'clock position.

Gerald Genta 1931–2011

Gérald Charles Genta was a Swiss watch designer and artist. He is known for specific timepieces like the Audemars Piguet Royal Oak, Cartier Pasha, and the Patek Philippe Nautilus, as well as his design work for IWC, Bulgari, Universal Genève and others.

DAY-DATE 36 SERIES (2000 TO 2007)

Rolex introduced the Day-Date 36 in the year 2000 with the six-digit 118000 series. Improvements were limited to the bracelet and clasp, but more dial options were available.

DAY-DATE II SERIES (2008 TO 2015)

The Day-Date II arrived in 2008, with a 41mm Oyster case and cal. 3156 (ref. 218238, 218239, 218235, 218206). In 2015, the reference was retired and replaced with the Day-Date 40.

DAY-DATE 40 SERIES

The Day-Date 40 Oyster case was reduced to 40mm and housed cal. 3255, which offered a longer power reserve.

STELLA DIALS

For a brief spell in the early 1970s, Rolex introduced some brightly colored enamel dials to the 36mm Day-Date. Thought to be named after either the American artist, Frank Stella, or Stella S.A., a Swiss paint manufacturer, they were targeted at the Middle Eastern market, were a commercial flop.

The paint used on these enamel dials was hand-mixed, so variations exist between batches of the same color. Enamel dials age better than plated and painted versions. Though prone to cracking and chipping if mistreated, their color tends to retain brightness and depth.

Today, these vintage 1970 models have become highly sought after as a unisex watch. In 2013, Rolex reintroduced some Day-Date Stella dials in blue, cherry, chocolate, rhodium, green, and cognac. These are only available on leather straps – not on the precious metal President bracelet, like the earliest Day-Date models.

Stella dials are a specialized domain. Brightly colored dials appearing on other references outside these date periods are most likely aftermarket fakes. Collectors should avoid these.

DAY-DATE REFERENCES

Name	Ref	Cal	Start	End	Bezel	Case	Description
Day-Date	**1800**	1555	1959	1987			
Day-Date	**1802**	1555	1959	1987		YG	
Day-Date	**1803**	1555	1959	1987	Fluted	YG	
Day-Date	**1804**	1555	1959	1987	Diamond	PL	
Day-Date	**1805**	1555	1959	1987	Diamond & Sapphire	YG	
Day-Date	**1806**	1555	1959	1987		YG	Morellis finish
Day-Date	**1807**	1555	1959	1987		YG	Bark finish
Day-Date	**1810**	1555	1959	1987		YG	
Day-Date	**1811**	1555	1959	1987		YG	
Day-Date	**1816**	1555	1959	1987	Baguette diamond		
Day-Date	**1817**	1555	1959	1987	Combination		
Day-Date	**1820**	1555	1959	1987			
Day-Date	**1823**	1555	1959	1987			
Day-Date	**1824**	1555	1959	1987			
Day-Date	**1829**	1555	1959	1987			
Day-Date	**1830**	1555	1959	1987			
Day-Date	**1831**					PL	
Day-Date	**1833**	1555	1959	1987			
Day-Date	**1834**	1555	1959	1987			
Day-Date	**1836**	1555	1959	1987			
Day-Date	**1837**	1555	1959	1987			
Day-Date	**1838**	1555	1959	1987			
Day-Date	**1839**	1555	1959	1987			
Day-Date	**1894**	1555	1959	1987			
Day-Date	**1895**	1555	1959	1987			
Day-Date	**6511**		1956	1959			
Day-Date	**6611**	1055	1957				
Day-Date	**6612**	1055	1957				
Day-Date	**6613**	1055	1957				
Day-Date	**17824**						
Day-Date	**18026**	3035					
Day-Date	**18028**	3035					
Day-Date	**18036**	3035					
Day-Date	**18038**	3035			Fluted	YG	
Day-Date	**18039**	3035					
Day-Date	**18046**	3035					
Day-Date	**18048**	3035					
Day-Date	**18049**	3035					
Day-Date	**18078**	3035					
Day-Date	**18079**	3035					
Day-Date	**18206**	3135	1988	2000			

SS : Stainless Steel | YG : Yellow Gold | RG : Rose Gold | WG : White Gold | PL : Platinum

Name	Ref	Cal	Start	End	Bezel	Case	Description
Day-Date	**18208**	3135	1988	2000			
Day-Date	**18238**	3135	1988	2000			
Day-Date	**18239**	3135	1988	2000			
Day-Date	**18248**	3135	1988	2000			
Day-Date	**18249**	3135	1988	2000			
Day-Date	**18296**	3135	1988	2000			
Day-Date	**18308**	3135					
Day-Date	**18338**	3135					
Day-Date	**18346**	3135					
Day-Date	**18348**	3135					
Day-Date	**18349**	3135					
Day-Date	**18366**	3135					
Day-Date	**18368**	3135					
Day-Date	**18378**	3135					
Day-Date	**18388**	3135					
Day-Date	**18389**	3135					
Day-Date	**18946**	3135					
Day-Date	**18948**	3135					
Day-Date	**18956**	3135					
Day-Date 36	**118205**	3155	2000	2007		RG	
Day-Date 36	**118206**	3155	2000	2007		PL	
Day-Date 36	**118208**	3155	2000	2007		YG	
Day-Date 36	**118209**	3155	2000	2007		WG	
Day-Date 36	**118235**	3155	2000	2007		RG	
Day-Date 36	**118238**	3155	2000	2007		YG	
Day-Date 36	**118239**	3155	2000	2007		WG	
Day-Date 36	**118296**	3155	2000	2007		PL	
Day-Date 36	**118338**	3155	2000	2007		YG	
Day-Date 36	**118339**	3155	2000	2007		WG	
Day-Date 36	**118346**	3155	2000	2007		PL	
Day-Date 36	**118348**	3155	2000	2007		YG	
Day-Date 36	**118366**	3155	2000	2007		PL	
Day-Date 36	**118388**	3155	2000	2007		YG	
Day-Date 36	**118389**	3155	2000	2007		WG	
Day-Date 36	**118398**	3155	2000	2007		YG	
Day-Date 40	**228235**	3255			Fluted	YG	
Day-Date 40	**228239**	3255			Fluted	PL	
Day-Date II	**216570**	3156	2008	2015			
Day-Date II	**218206**	3156	2008	2015		PL	
Day-Date II	**218235**	3156	2008	2015		RG	
Day-Date II	**218238**	3156	2008	2015		YG	
Day-Date II	**218239**	3156	2008	2015		WG	

ROLEX BUBBLEBACK

The Bubbleback is an unofficial category of a Rolex watch made between the mid-1930s and the mid-1950s. Numerous references fall into the Bubbleback category including several semi-Bubblebacks and transitional examples.

The term Bubbleback comes from a pronounced rounded and protruding caseback. These watches also go by the nickname, "ovettone" which is Italian for little egg. This dome-shaped caseback was necessary to accommodate the rotor of the early self-winding movements. The design choice for a convex caseback, as opposed to a thicker mid case (and flat caseback), resulted in a distinctive and unusual look, which proved very comfortable on the wrist.

Bubbleback case diameters are small – typically 30mm to 36mm for men's models. Combined with a domed acrylic crystal the convex casebacks contributed to an almost egg-like profile, standing high and proud on the wrist. The look is certainly distinctive and immediately recognizable. The cases of later references are 36mm and are known as the Big Bubbleback. These are the precursors to the Datejust.

The first Bubbleback was the ref. 1858, with a caliber 520 movement and a three-piece case design. In 1936, the two-piece case was introduced with refs 3131 and 3132.

In recent decades, the Bubbleback has fallen out of fashion, as tastes now tend toward larger watch sizes. However, they represent an early and vital phase in Rolex's history and continue to be of great interest to collectors.

These self-winding movements and early Oyster cases are the precursors to the modern Oyster Perpetual. Nearly every contemporary Rolex now features the word "Perpetual" on its dial. In this sense, the Bubbleback is historically significant, and its role and place in the evolution of the Rolex product line cannot be understated.

BUBBLEBACK CALIFORNIA DIALS

Bubblebacks are the only models to use the California Dial. These dials consist of half Roman and half Arabic numerals. Numbers 10 to 2 are Roman, 4 to 8 are Arabic. Rolex and Panerai are the two brands most commonly associated with the California dial. The first Rolex Bubbleback to appear with such a dial was the ref. 3595.

The origins of the name are unclear, but authoritative sources such as James Dowling claim it comes from a Californian dial refinisher, Kirk Rich, in the 1970s. Kirk became so well known for his high-quality restoration of these dials that clients began to call them the 'California' dial.

BUBBLEBACK REFERENCES

Ref	Cal	Start	Bezel	Size	Case	Description
1858/0	520		Smooth (Polished)		SS	Perpetual, Chronometer, Seconds sub-dial
1858/1	520		Smooth (Polished)		YGF	Perpetual, Chronometer, SS back
1858/3	520		Smooth (Polished)		SSRG	Perpetual, Chronometer
1858/7	520		Smooth (Polished)		RG	Perpetual, Chronometer
1858/7	520		Smooth (Polished)		YG	Perpetual, Chronometer
1858/7	520		Smooth (Polished)		RG	Perpetual, Chronometer
1858/8	520		Smooth (Polished)		RG	Perpetual, Chronometer
1873/0	8¾		Smooth (Polished)		SS	Perpetual, Chronometer
1873/1	Chronometer		Smooth (Polished)		YGF	Perpetual, SS back
1873/3	Chronometer		Smooth (Polished)		SSYG	Perpetual
1873/7	Chronometer		Smooth (Polished)		RG	Perpetual
1873/7	Chronometer		Smooth (Polished)		YG	Perpetual
1873/7	Chronometer		Smooth (Polished)		RG	Perpetual
1873/8	Chronometer		Smooth (Polished)		RG	Perpetual
2240/0	9¾		Smooth (Polished)		SS	Perpetual, Dennison case
2280/0	Manual		Smooth (Polished)		SS	"Rolco"
2764/0	Chronometer		Engine Turned		SS	Perpetual, Seconds sub-dial
2764/1	Chronometer		Smooth (Polished)		YGF	Perpetual, SS back
2764/3	Chronometer		Smooth (Polished)		SSYG	Perpetual
2764/7	Chronometer		Smooth (Polished)		YG	Perpetual
2764/7	Chronometer		Smooth (Polished)		RG	Perpetual
2764/7	Chronometer		Smooth (Polished)		YG	Perpetual
2764/7	Chronometer		Smooth (Polished)		RG	Perpetual
2764/8	Chronometer		Smooth (Polished)		YG	Perpetual
2764/8	Chronometer		Smooth (Polished)		RG	Perpetual
2765/0	Extra Prima		Smooth (Polished)			
2765/3					SSYG	
2784/0	Manual		Smooth (Polished)			"Junior Sport", Seconds sub-dial
2940/0	Super Precision		Smooth (Polished)		SS	Perpetual
2940/3	Chronometer		Engine Turned		SSYG	Perpetual, Seconds sub-dial
2940/5	Chronometer		Engine Turned		YGF	Perpetual, SS back
2940/7	Chronometer		Engine Turned		RG	Perpetual
2945/0	Chronometer		Smooth (Polished)		SS	Perpetual
3042/3	Manual		Engine Turned		SSRG	
3065/0	Chronometer		Smooth (Polished)		SS	Perpetual, Hooded lugs

SS : Stainless Steel | YG : Yellow Gold | RG : Rose Gold | WG : White Gold

BUBBLEBACK REFERENCES CONT.

Ref	Cal	Start	Bezel	Size	Case	Description
3065/3		1939	Smooth (Polished)		SSRG	Perpetual, Hooded lugs
3065/7			Smooth (Polished)		YG	Perpetual, Hooded lugs
3065/7			Engine Turned		YG	Perpetual, Hooded lugs
3065/7			Smooth (Polished)		RG	Perpetual, Hooded lugs
3065/8			Smooth (Polished)		RG	Perpetual, Hooded lugs
3130/7	Chronometer		Smooth (polished)		YG	Perpetual, Seconds sub-dial
3130/7	Chronometer		Smooth (polished)		RG	Perpetual, Seconds sub-dial
3130/8	Chronometer		Smooth (polished)		YG	Perpetual, Seconds sub-dial
3130/8	Chronometer		Smooth (polished)		RG	Perpetual, Seconds sub-dial
3131/7	620	1936	Engine Turned		RG	Perpetual, Chronometer
3131/7	620	1936	Smooth (Polished)		RG	Perpetual, Chronometer
3131/7	620	1936	Smooth (Polished)		YG	Perpetual, Chronometer
3131/8	620	1936	Smooth (Polished)		YG	Perpetual, Chronometer
3131/8	620	1936	Smooth (Polished)		RG	Perpetual, Chronometer
3132/0	630	1936	Smooth (Polished)		SS	Perpetual, Chronometer
3132/3	630	1936	Engine Turned		SSRG	Perpetual, Chronometer, Seconds sub-dial
3132/3	630	1936	Engine Turned		SSYG	Perpetual, Chronometer, Seconds sub-dial
3132/7	630	1936	Engine Turned			Perpetual, Chronometer, Seconds sub-dial
3132/7	630	1936	Engine Turned		RG	Perpetual, Chronometer, Seconds sub-dial
3132/7	630	1936	Engine Turned		YG	Perpetual, Chronometer, Seconds sub-dial
3132/7	630	1936	Engine Turned		RG	Perpetual, Chronometer, Seconds sub-dial
3132/8	630	1936	Engine Turned		YG	Perpetual, Chronometer, Seconds sub-dial
3133/0	Chronometer		Smooth (Polished)		SS	Perpetual
3133/3	Chronometer		Engine Turned		SSYG	Perpetual
3133/3	Chronometer		Smooth (Polished)		SSRG	Perpetual
3133/7	Chronometer		Smooth (Polished)			Perpetual
3133/7	Chronometer		Smooth (Polished)		RG	Perpetual
3133/7	Chronometer		Smooth (Polished)			Perpetual
3133/7	Chronometer		Smooth (Polished)		YG	Perpetual
3133/7	Chronometer		Smooth (Polished)		RG	Perpetual
3133/8	Chronometer		Smooth (Polished)		YG	Perpetual
3133/8	Chronometer		Smooth (Polished)		RG	Perpetual
3134/0	Chronometer		Smooth (Polished)		SS	Perpetual
3134/1	Chronometer		Smooth (Polished)		YGF	Perpetual, SS back
3134/3	Chronometer		Smooth (Polished)		SSYG	Perpetual
3134/7	Chronometer		Smooth (Polished)		YG	Perpetual

Ref	Cal	Start	Bezel	Size	Case	Description
3134/7	Chronometer		Smooth (Polished)		RG	Perpetual
3134/7	Chronometer		Smooth (Polished)		YG	Perpetual
3134/7	Chronometer		Smooth (Polished)		RG	Perpetual
3134/8	Chronometer		Smooth (Polished)		RG	Perpetual
3134/8	Chronometer		Smooth (Polished)		YG	Perpetual
3135/0	Chronometer		Smooth (Polished)			Perpetual
3136/0	Manual		Smooth (Polished)	Mid Size	SS	"Junior Sport"
3333/3			Smooth (Polished)		SSYG	Perpetual, Hooded lugs
3348/0			Engine Turned	Mid Size	SS	Perpetual, Seconds sub-dial
3353/0			Smooth (Polished)	Mid Size	SS	Perpetual, Hooded lugs
3353/3			Smooth (Polished)		SSYG	Perpetual, Hooded lugs
3353/8			Smooth (Polished)		YG	Perpetual, Hooded lugs
3358/0			Fluted	Mid Size	SS	Perpetual
3372/0	630		Smooth (Polished)		SS	Perpetual, Chronomter
3372/2	630				RG	Perpetual, Chronomter
3372/3	630				SS	Perpetual, Chronometer, Hooded lugs
3372/3	630				SSRG	Perpetual, Chronometer, Seconds sub-dial
3372/7	630				YG	Perpetual, Chronomter
3372/7	630				RG	Perpetual, Chronomter
3372/8	630				YG	Perpetual, Chronomter
3372/8	630				RG	Perpetual, Chronomter
3548/0	Chronometer		Smooth (Polished)		SS	Perpetual
3548/8	Chronometer		Smooth (Polished)		YG	Perpetual
3549/0	Chronometer		Smooth (Polished)		SS	Perpetual
3595/3	Chronometer		Smooth (Polished)		SSYG	Perpetual, Seconds sub-dial, California dials
3595/3	Chronometer		Smooth (Polished)		SSRG	Perpetual, Seconds sub-dial, California dials
3598/3			Smooth (Polished)		SS	Perpetual, Seconds sub-dial
3599/0			Smooth (Polished)		SS	Perpetual, Hooded lugs
3696/3	Chronometer		Smooth (Polished)		SSRG	Perpetual
3725/3			Engine Turned	32mm	SSYG	Perpetual
3725/7			Engine Turned	32mm	RG	Perpetual, Seconds sub-dial
3725/8			Engine Turned	32mm	RG	Perpetual
3767/8	Chronometer		Smooth (Polished)	Boys Size	RG	Perpetual
3795/8			Engine Turned		RG	Perpetual, Seconds sub-dial
4392/0	Chronometer		Smooth (Polished)		SS	Perpetual

SS : Stainless Steel | YG : Yellow Gold | RG : Rose Gold | WG : White Gold

BUBBLEBACK REFERENCES CONT.

Ref	Cal	Start	Bezel	Size	Case	Description
4392/7	Chronometer		Smooth (Polished)		YG	Perpetual
4392/8	Chronometer		Smooth (Polished)		RG	Perpetual
4453/0	Manual		Smooth (Polished)	Mid Size	SS	"Junior Sport"
4486/8			Smooth (Polished)	Ladies	YG	Perpetual, Seconds sub-dial
4486/8			Smooth (Polished)	Ladies	RG	Perpetual, Seconds sub-dial
4777/8	Chronometer		Engine Turned		YG	Perpetual
4919/3	Chronometer		Engine Turned		SSYG	Perpetual
4939/3	Chronometer		Smooth (Polished)		SSYG	Perpetual, Seconds sub-dial
4961/0	Chronometer				SS	Perpetual, Cushion case
5001/8	Chronometer		Engine Turned		YG	Perpetual, Seconds sub-dial
5002/7			Smooth (Polished)	Ladies	YG	Perpetual, Seconds sub-dial
5002/7			Smooth (Polished)	Ladies	YG	Perpetual, Seconds sub-dial
5002/8			Engine Turned	Ladies	RG	Perpetual, Seconds sub-dial
5003/3	Super Precision		Engine Turned	Ladies	SSYG	Perpetual
5003/7			Smooth (Polished)	Ladies	YG	Perpetual
5003/8			Engine Turned	Ladies	RG	Perpetual
5003/8			Engine Turned	Ladies	YG	Perpetual
5006/3	Chronometer		Smooth (Polished)	Mid Size	SSYG	Perpetual
5006/7	Chronometer		Smooth (Polished)	Mid Size	YG	Perpetual
5007/3	Chronometer		Smooth (Polished)		SS 4Y	Perpetual
5010/0	Chronometer		Smooth (Polished)		SS	Perpetual
5010/3	Chronometer		Smooth (Polished)		SSYG	Perpetual
5010/3	Chronometer		Smooth (Polished)		SSRG	Perpetual
5011/3	Chronometer		Engine Turned		SSYG	Perpetual, SS back
5011/3	Super Precision		Engine Turned		SSRG	Perpetual
5011/7	Chronometer		Engine Turned		RG	Perpetual
5011/8	Chronometer		Engine Turned		YG	Perpetual
5013/3	Chronometer		Engine Turned		SSYG	Perpetual, Seconds sub-dial
5015/0	Chronometer		Smooth (Polished)		SS	Perpetual
5015/3	Chronometer	1948	Engine Turned		SSRG	Perpetual
5015/3	Chronometer		Engine Turned		YG	Perpetual
5015/3	Chronometer		Engine Turned		RG	Perpetual
5015/8	Chronometer		Engine Turned		YG	Perpetual
5015/8	Chronometer		Engine Turned		RG	Perpetual
5026/0			Smooth (Polished)			Perpetual, Seconds sub-dial
5045/3	Chronometer		Engine Turned		SSYG	Perpetual

Ref	Cal	Start	Bezel	Size	Case	Description
5048/0	Chronometer		Smooth (Polished)		SS	Perpetual
5048/3	Chronometer		Smooth (Polished)			Perpetual
5048/8	Chronometer		Smooth (Polished)		YG	Perpetual
5050/0	Chronometer		Smooth (Polished)		SS	Perpetual
5050/7	Chronometer		Smooth (Polished)		YG	Perpetual
5050/8	Chronometer		Smooth (Polished)		YG	Perpetual
5051/7	Chronometer		Smooth (Polished)		RG	Perpetual
5055/3	Chronometer		Smooth (Polished)	Mid Size	SSYG	Perpetual
5105/0	Chronometer		Engine Turned		SS	Perpetual
5105/1	Chronometer		Engine Turned		YGF	Perpetual, SS back
5105/3	Chronometer		Engine Turned		SSYG	Perpetual
5105/7	Chronometer		Smooth (Polished		YG	Perpetual
5105/7	Chronometer		Engine Turned		RG	Perpetual
5105/7	Chronometer		Engine Turned		YG	Perpetual
5105/7	Chronometer		Engine Turned		RG	Perpetual
5105/8	Chronometer		Engine Turned		YG	Perpetual
5105/8	Chronometer		Engine Turned		RG	Perpetual
6006/0			Smooth (Polished)	Mid Size	SS	Perpetual, Seconds sub-dial
6006/8			Smooth (Polished)	Mid Size	RG	Perpetual, Seconds sub-dial
6015/0	Chronometer		Engine Turned		SS	Perpetual
6048/8	Chronometer		Smooth (Polished)		RG	Perpetual
6084/0	645		Smooth (Polished)	Full Size	SS	Perpetual, Chronometer
6106/0	Chronometer		Smooth (Polished)		SS	Perpetual
6428/0	Chronometer				SS	Perpetual
8056/3	Chronometer		Smooth (Polished)		SSYG	Perpetual, Hooded lugs
8056/3	Chronometer		Smooth (Polished)		SSRG	Perpetual, Hooded lugs
8056/7	Chronometer		Smooth (Polished)		YG	Perpetual, Hooded lugs
8056/7	Chronometer		Smooth (Polished)		RG	Perpetual, Hooded lugs
8056/8	Chronometer		Smooth (Polished)		YG	Perpetual, Hooded lugs
8056/8	Chronometer		Smooth (Polished)		RG	Perpetual, Hooded lugs

SS : Stainless Steel | YG : Yellow Gold | RG : Rose Gold | WG : White Gold

Speleology | ˌspē-lē-ä-lə-jē |

noun ***scientific, geography***

- the scientific study or exploration of caves.
- from Greek spēlaion, "cave, cavern".
- the study of cave geological formation, flora, fauna.
- the sport or pastime of exploring caves.

ROLEX
20 WORLD'S RECORDS
GREAT BRITAIN PATENTS
260554/1925
274.789
SWISS PATENTS
+114.948
+120.851

Chapter 09
FORMAL DRESS

When Andre Heiniger was appointed the CEO of Rolex in 1963, he made a conscious decision to reinvent Rolex as a luxury brand and move it upmarket. This change of strategy was a marked departure from his predecessor's focus on high-specification, professional tool watches.

Rolex had launched several dress watches before the arrival of the now-famous Cellini collection. The Rolex Prince Brancard of 1928 is an iconic example. While it came to be known as The Doctor's Watch, it was not explicitly designed as such, and would not have been considered a dress watch by Heiniger.

> "Fashion is ephemeral, dangerous and unfair"
>
> *Karl Lagerfeld*

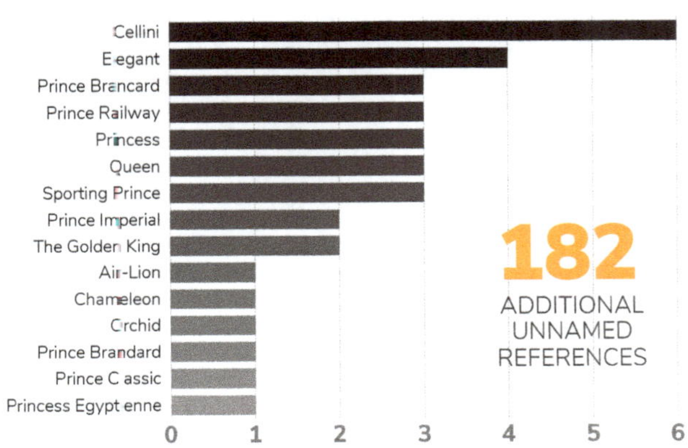

182 ADDITIONAL UNNAMED REFERENCES

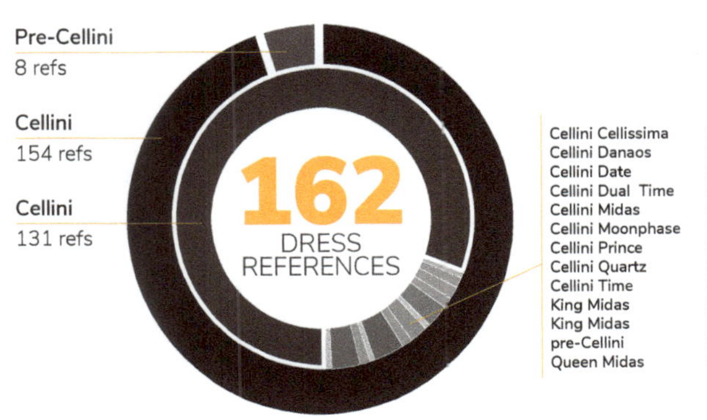

The Cellini collection emerged in the late 1960s under Heiniger's guidance and was named after Benevento Cellini (1500-1571), an artist and adventurer. Cellini served as a goldsmith and sculptor to popes and royalty during the Italian Renaissance.

The Rolex Cellini collection is thought to be the first luxury watch produced in large quantities to be awarded chronometric certification. The understated dials of vintage pieces display only time, with no second hand. Modern Cellini watches have adopted complications to make them more practical for daily wear. Cellinis are described as water resistant but are not waterproof. They can withstand a rainstorm or hand washing but are not suitable for the hot tub or swimming pool.

The Cellini collection falls into the Modern Classic taxonomy, though there is disagreement over how classic any of these designs are. Like ladies' vintage Rolex references in general, Cellini dress watches have a small but enthusiastic following.

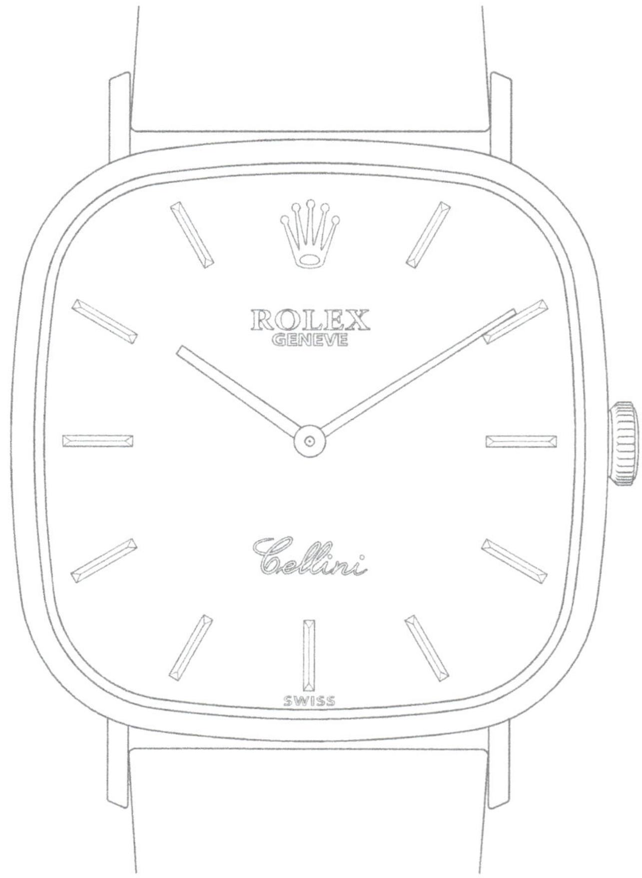

Benvenuto Cellini (1500 - 1571)

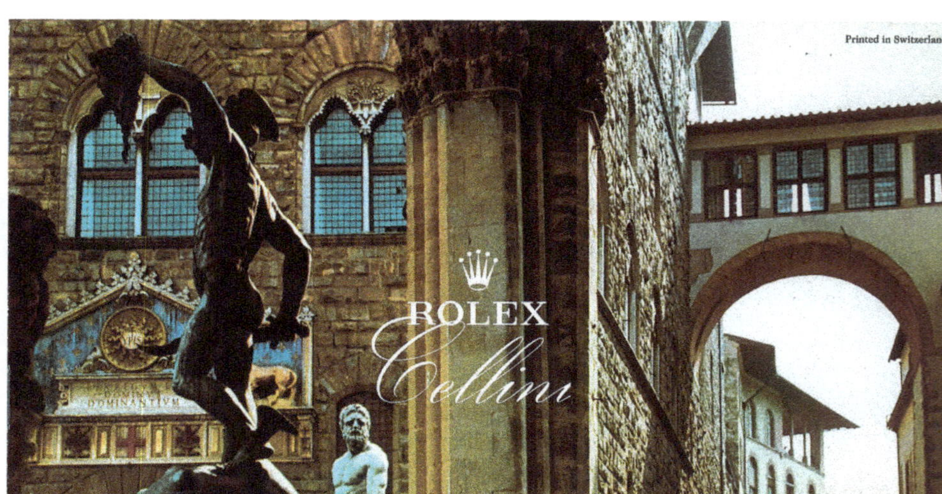

The beautifully designed Cellini handbook from 2007, prominently featuring the bronze statue of Perseus.

CELLINI CESTELLO (1990 TO 2000)

The name Cestello was used in the 1990s to describe a series of round dress watches offered in platinum, yellow, or rose gold. It appeared to be an attempt to create a product line within the Cellini Collection. The name was used in catalogs and advertising but did not appear on dials. The name and watches failed to distinguish themselves as a commercial success and were quietly retired.

CELLINI DANAOS (2000)

The Danaos line appeared in 2000. There is some disagreement about the origin of the name. Some sources claim it is a Latin reference to Greeks, and others claim it refers to an Egyptian king or the mythological king of the Mycenaean city of Argos.

The Danaos is available in precious metals and two-tone combinations on leather straps. It is a distinctive and contemporary looking line of watches. At 39mm the pieces had more presence than the rather bland Cestello references they replaced. The Danaos references went on to have moderate commercial success, and prices on the pre-owned market are holding up well.

KING MIDAS

The designer of the Rolex King Midas was Gérald Genta, the most famous watch designer in modern times. Gérald Genta was also responsible for the Audemars Piguet Royal Oak and the Patek Philippe Nautilus.

The King Midas has an asymmetric pentagonal case inspired by the Greek Parthenon. In Greek mythology, King Midas could turn anything he touched to gold.

In the late 1960s, the first King Midas, ref. 9630, debuted and was limited to only 1,000 numbered units. This original version pre-dated the Cellini collection, but later versions became part of the group and known as Cellini King Midas.

At launch, it was the most opulent and expensive watch Rolex offered. They are stamped from a solid block of 18K yellow gold and weigh 150 to 200 grams. The considerable weight comes from the integrated bracelet whichbracelet, which is characteristic of many Gerald Genta watches.

The winding crown of ref. 9630 does not feature a Rolex coronet like later versions, but instead, has one shaped like the sun. It is positioned on the left side of the case – a further reference to King Midas, whose left hand had the mythical golden touch. The shape and position resemble the sun rising over the Parthenon. Engraved on either side of the left-handed crown are the words, "King Midas", framing the rising sun.

Like most dress watches it has only an hour and minute hand. The dial features the Rolex coronet at 12 o'clock and "Midas" written in Greek below the centerline. These Greek letters mark the ref. 9360 as unique. Later editions from the ref. 3580 onwards feature the name Cellini instead.

Famous owners of the 9360 King Midas include Elvis Presley and John Wayne. If the King Midas appeals to you, the original and limited edition 9630 is the one to own.

Ref. 3580 arrived in 1974 as a white gold version of the original 9360. A very rare Queen Midas followed shortly after. They were also individually numbered but made in larger volume.

CELLINI PRINCE

The Rolex Cellini Prince is a reinterpretation of the original Rolex Prince of 1928. The modern reinterpretation debuted in 2005 and has a distinct Art Deco aesthetic reminiscent of the roaring 1920s.

The original Prince had a distinctive avant-garde rectangular case. The modern version has a similar rectangular case and is equally striking. One significant difference is the display case back to show off the highly finished, manual wind caliber 7040.

The Rolex Cellini Prince (refs. 54425, 54419 and 54439) is made only in precious metal cases (18K yellow gold, white gold, and Everose gold) and five dial combinations. If you like the dressier, Art Deco aesthetic, these are very attractive, well finished and sought after pieces.

Silver dial with a godron circulaire. This pattern is a radial or circular, striped ring design

Diamond-pace dial with solver godron circulaire motif

Champagne dial decorated with a clou de Paris or guilloch motif. This design is a hobnail or checkered pattern

Black and pink dial with rayon flamme de la gloire

Black and silver dial with double rayon flamme de la gloire. This pattern is a radiating striped design, which appears to originate from the center of the dial

DRESS WATCH REFERENCES

Name	Ref	Cal	Size	Description
Cellini	**2466**	Manual	Ladies	Round
Cellini	**2704**	Manual	Unisex	Square
Cellini	**3224**	Manual	Full Size	Square
Cellini	**3612**	650	Unisex	Round
Cellini	**3718**		Full Size	Round
Cellini	**3735**	Manual	Full Size	Square
Cellini	**3759**		Full Size	Round
Cellini	**3761**	Manual	Unisex	Round
Cellini	**3783**			
Cellini	**3787**	Manual	Unisex	Round
Cellini	**3804**		Full Size	Round
Cellini	**3811**	1600	Full Size	Square
Cellini	**3833**	Manual	Full Size	Round
Cellini	**3834**	Manual	Full Size	Rectangular
Cellini	**4014**	1600	Full Size	Rectangular
Cellini	**4016**	Manual	Full Size	Rectangular
Cellini	**4032**		Unisex	Rectangular
Cellini	**4041**		Unisex	Round
Cellini	**4043**		Unisex	Oval
Cellini	**4080**		Unisex	Square
Cellini	**4081**	Manual	Ladies	Round
Cellini	**4082**		Ladies	Square
Cellini	**4083**		Full Size	Round
Cellini	**4084**		Full Size	Square
Cellini	**4087**		Unisex	Square
Cellini	**4100**		Full Size	Rectangular
Cellini	**4101**		Full Size	Rectangular
Cellini	**4105**	1601	Full Size	Other
Cellini	**4106**		Unisex	Hexagonal (6)
Cellini	**4107**		Unisex	Hexagonal (6)
Cellini	**4108**		Unisex	Rectangular
Cellini	**4109**		Full Size	Round
Cellini	**4110**		Unisex	Oval
Cellini	**4111**		Unisex	Oval
Cellini	**4112**	1601	Full Size	Round
Cellini	**4113**		Unisex	Other
Cellini	**4114**		Unisex	Square
Cellini	**4121**	1601	Full Size	Hexagonal (6)
Cellini	**4122**	Manual	Full Size	Hexagonal (6)
Cellini	**4126**	1601	Full Size	Square
Cellini	**4127**	Manual	Unisex	Rectangular
Cellini	**4129**		Unisex	Round
Cellini	**4131**		Unisex	Rectangular
Cellini	**4132**		Unisex	Rectangular
Cellini	**4133**	Manual	Full Size	Round
Cellini	**4135**		Unisex	Square
Cellini	**4136**		Unisex	Hexagonal (6)
Cellini	**4139**		Ladies	Rectangular
Cellini	**4140**			
Cellini	**4143**			
Cellini	**4211**			
Cellini	**4302**		Ladies	Rectangular
Cellini	**4304**		Ladies	Round
Cellini	**4305**			
Cellini	**4306**		Full Size	Round
Cellini	**4307**			
Cellini	**4308**			
Cellini	**4310**		Full Size	Rectangular
Cellini	**4312**			
Cellini	**4317**		Unisex	Round
Cellini	**4318**		Unisex	Square
Cellini	**4319**		Unisex	Round
Cellini	**4320**		Unisex	Square
Cellini	**4321**	Manual	Ladies	Rectangular
Cellini	**4322**	Manual	Full Size	Other
Cellini	**4324**	Manual	Full Size	Hexagonal (6)
Cellini	**4327**	Manual	Full Size	Square
Cellini	**4328**		Full Size	Square
Cellini	**4329**			
Cellini	**4331**			
Cellini	**4332**	Manual	Ladies	Square
Cellini	**4333**	Manual	Unisex	Square
Cellini	**4335**	Manual	Ladies	Square
Cellini	**4339**	Manual	Ladies	Rectangular
Cellini	**4340**			
Cellini	**4341**	Manual	Ladies	Hexagonal (6)
Cellini	**4343**			
Cellini	**4344**	Manual	Full Size	Other
Cellini	**4347**			
Cellini	**4349**	Manual	Full Size	Other

Name	Ref	Cal	Size	Description
Cellini	**4350**	Manual	Full Size	Other
Cellini	**4378**			
Cellini	**4379**	Manual	Full Size	Round
Cellini	**4614**	1601	Ladies	Oval
Cellini	**4615**			
Cellini	**4621**			
Cellini	**4622**			
Cellini	**4624**		Unisex	Round
Cellini	**4625**	Manual	Ladies	Oval
Cellini	**4626**			
Cellini	**4628**			
Cellini	**4629**			
Cellini	**4630**		Unisex	Round
Cellini	**4631**			
Cellini	**4632**			
Cellini	**4633**			
Cellini	**4636**			
Cellini	**4650**		Full Size	Other
Cellini	**4652**		Full Size	Other
Cellini	**4942**	Manual	Ladies	Other
Cellini	**4943**	Manual	Ladies	Other
Cellini	**5109**	1601	Ladies	Round
Cellini	**5113**		Unisex	Round
Cellini	**5114**		Unisex	Round
Cellini	**5116**	1602	Full Size	Round
Cellini	**5156**	1601	Full Size	Other
Cellini	**5166**		Full Size	Round
Cellini	**5167**			
Cellini	**5171**	Manual	Ladies	Round
Cellini	**5172**			
Cellini	**5184**	Manual	Ladies	Round
Cellini	**5188**		Ladies	Round
Cellini	**5191**			
Cellini	**5192**			
Cellini	**5221**	Manual	Ladies	Oval
Cellini	**5222**			
Cellini	**5241**	Manual	Full Size	Round
Cellini	**5330**	Manual	Full Size	Other
Cellini	**5443**	Manual	Full Size	Rectangular
Cellini	**6110**	A260	Ladies	Round

Name	Ref	Cal	Size	Description
Cellini	**6628**	1135		
Cellini	**6673**	6620	Ladies	Round
Cellini	**6692**		Ladies	Other
Cellini	**14233M**			
Cellini Cellissima	**6693**	1215	Ladies	Other
Cellini Danaos	**4223**			
Cellini Danaos	**4243**	Manual	Full Size	Round
Cellini Date	**50515**	3135	Full Size	Round
Cellini Date	**50519**	3165	Full Size	Round
Cellini Dual Time	**50525**	3135	Full Size	Round
Cellini Dual Time	**50529**	3180	Full Size	Round
Cellini Midas	**4294**		Unisex	Pentagonal
Cellini Midas	**4315**		Unisex	Pentagonal
Cellini Midas	**4336**		Unisex	Pentagonal
Cellini Midas	**4342**		Full Size	Pentagonal
Cellini Moonphase	**50535**	3165	Full Size	Round
Cellini Prince	**54419**	7040	Full Size	Rectangular
Cellini Prince	**54425**	7040		
Cellini Prince	**54439**	7040	Full Size	Rectangular
Cellini Quartz	**6621**	1130	Ladies	Round
Cellini Quartz	**6623**	6620	Full Size	Round
Cellini Time	**50505**	3132	Full Size	Round
Cellini Time	**50509**	3132	Full Size	Round
Cellini Time	**50609**	3132	Full Size	Round
Cellini Time	**50705**	3132	Full Size	Round
Cellini Time	**50709**	3132	Unisex	Round
King Midas	**3580**		28mm	Pentagonal
King Midas	**4015**	1601	28mm	Pentagonal
King Midas	**9630**	650	28mm	Pentagonal First, original King Midas
Pre-Cellini	**2736**	Manual	Full Size	Square
Pre-Cellini	**3546**	1400	Ladies	Round
Pre-Cellini	**3737**	Manual	Full Size	Square
Pre-Cellini	**4102**		Full Size	Rectangular
Pre-Cellini	**4103**		Unisex	Rectangular
Pre-Cellini	**4104**		Full Size	Rectangular
Queen Midas	**3581**		Ladies	Pentagonal

ROLEX
20 WORLD'S RECORDS
GREAT BRITAIN PATENTS
260554/1925
274.789
SWISS PATENTS
+114.948
+120.851

1929

Parthenon | pahr-thuh-nän |

noun ***historical Greek***

- the temple of Athena Parthenos on the Acropolis at Athens, competed c438 b.c.
- regarded as the finest Doric temple.
- inspiration for the rare Rolex King Midas reference 9630.

Chapter 10

BRACELETS

Until the era of Modern Classics, Rolex had been using Oyster and Jubilee bracelets from the bracelet-maker Gay Frères. This company is also known for the famous Heuer beads of rice bracelet and the bracelet for the Audemars Piguet Royal Oak.

Rolex used Gay Frères exclusively until the late 1940s. From the 1950s to the mid-1970s Rolex used several suppliers to create the same bracelet but for different territories. For example, Rolex used suppliers in North America in response to US tax regulations.

C&I supplied the USA-made riveted Oyster bracelets. There are also Rolex Oyster and Jubilee bracelets marked “Hecho en Mexico” on the clasp, indicating production in Mexico.

Rolex acquired the Gay Frères company in 1998, bringing all of their bracelet production in-house.

Rolex introduced their iconic Jubilee bracelet with the Datejust in 1945/6, at a time when bracelets were considered more decorative than functional. The following year Rolex was awarded a patent for the Oyster bracelet (1947), which appeared in the Rolex catalog in 1948. Throughout the 1940s, Rolex did not offer a bracelet as standard, but as an expensive upgrade. Leather straps were the most common way of wearing a watch. They were treated as consumable and disposable – particularly in hot and humid climates where they would degrade quickly.

There were a great many bracelet styles produced, but the two most commonly recognized are the Oyster and Jubilee.

“There are two kinds of fools: one says, “This is old, therefore it is good”; the other says, “This is new, therefore it is better.”

Dean William Ralph Inge

OYSTER BRACELET

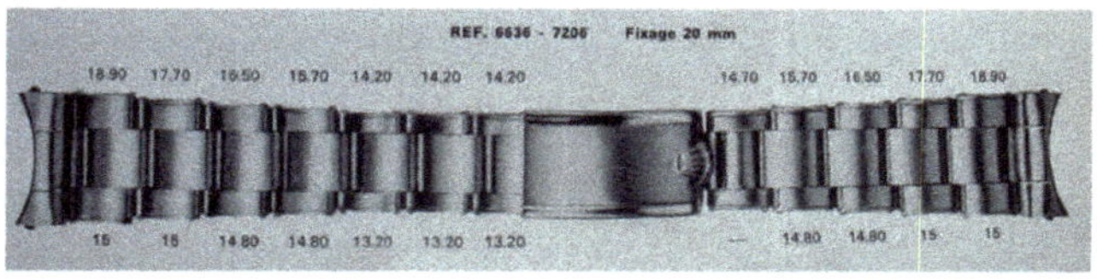

The first generation from the 1950s was the rivet style bracelet with visible rivet studs on the outer edge of the hollow folded links. Links were fastened together with the rivets having a peaked, conical cap that became less pronounced with polishing. These bracelets were also offered in an expanding link style (refs. 6634, 6635, 6636), but this was phased out after proving less resilient and less comfortable.

The second generation (the 1960s) was the folded link style bracelet (refs. 7834, 7835, 7836, 9315). The folded link, as the name suggests is made by folding sheet metal in on itself multiple times. These Swiss rolls result in a thicker link, but the pins holding them together are hidden.

The last and current generation is the solid link bracelet. These used the same numbering convention as the second generation, but with an added zero suffix. (e.g. 93150). The solid link style has proven the most robust, with even end links evolving from folded steel to solid. These are made in either a fully-brushed or partially polished (center link) finish.

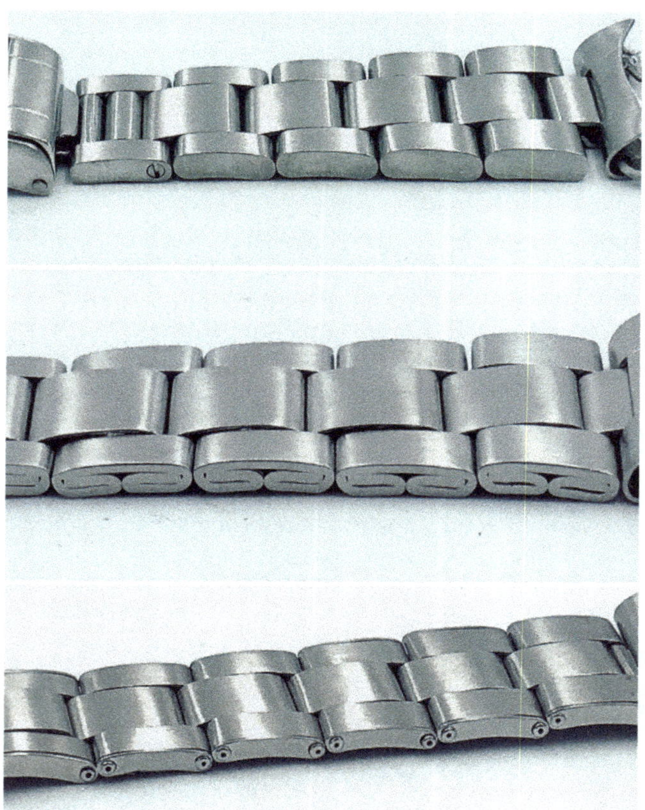

Top: Folded links (1960s)
Middle: Rivetted links (1950s)
Bottom: Solid links (1980s)

The Oyster bracelet reference number is usually found on the first or final link of the bracelet. This code indicates which generation it belongs to (four-digit or five-digit) and the appropriate end link. For example, the ref. 7206 is a riveted Oyster requiring 20mm end links, while the ref. 7205 is the very same bracelet but requires 19mm end links.

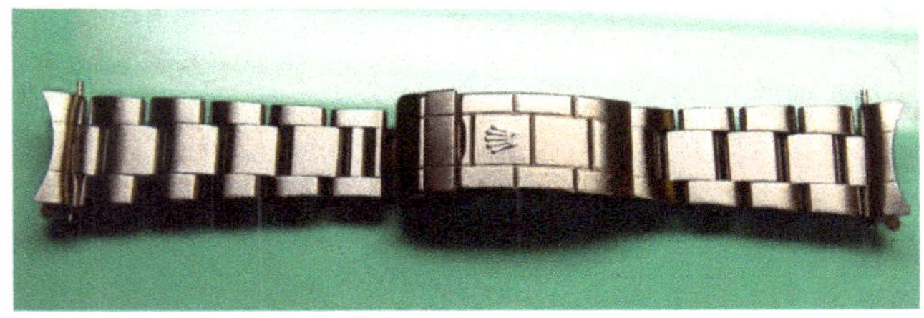

Getting a comfortable fit with an Oyster bracelet and Fliplock clasp can be challenging for some wrist shapes. Half links are available as an option.

Removing a permanent link is possible, and some independent watchmakers or jewelers may be willing to perform this procedure. While an AD won't perform this, a pair of pliers and masking tape can do the job. The process is non-reversible, damages the permanent link during removal, and should be considered a last resort. Correcting an Oyster that has had this procedure, can be done by adding back a removable link, however not all the permanent links are the same size so such a correction may fix the length but leave a mismatched width.

This is a Service Replacement clasp denoted by the 'S'. The AB3 code suggests production in the third quarter of the year 2000. The 93150 denotes an Oysterlock clasp for a Submariner or Sea-Dweller and should be paired with a matching Oyster bracelet marked 93150.

JUBILEE BRACELET

The Jubilee bracelet arrived in 1945 with the new Datejust and was later introduced as an option on the GMT-Master and the early Cosmograph Daytona. The Jubilee should not be confused with the flagship President bracelet.

Each link of the Jubilee is made of five-pieces, comprising three thinner inner links flanked by larger outer links. The two different link sizes are most noticeable on two-tone Rolesor versions with the inner links in yellow or Everose gold.

The Jubilee bracelet can be fitted with a concealed folding Crownclasp, which has a Rolex coronet lever to open the bracelet and reveal the folding blades. The concealment of the clasp allows the pattern of the Jubilee links to run seamlessly around the wrist.

The Jubilee bracelet is prone to wear and tear. Each of the five links tend toward friction-wear, resulting in a rounding of corners and edges and the internal retaining pin is also prone to bending. This results in an overall loosening of the bracelet described as stretching or sloppiness. Depending on the degree of wear and tear, Jubilee bracelets can be restored by replacing the pins. If the links are seriously worn, laser welding may be needed.

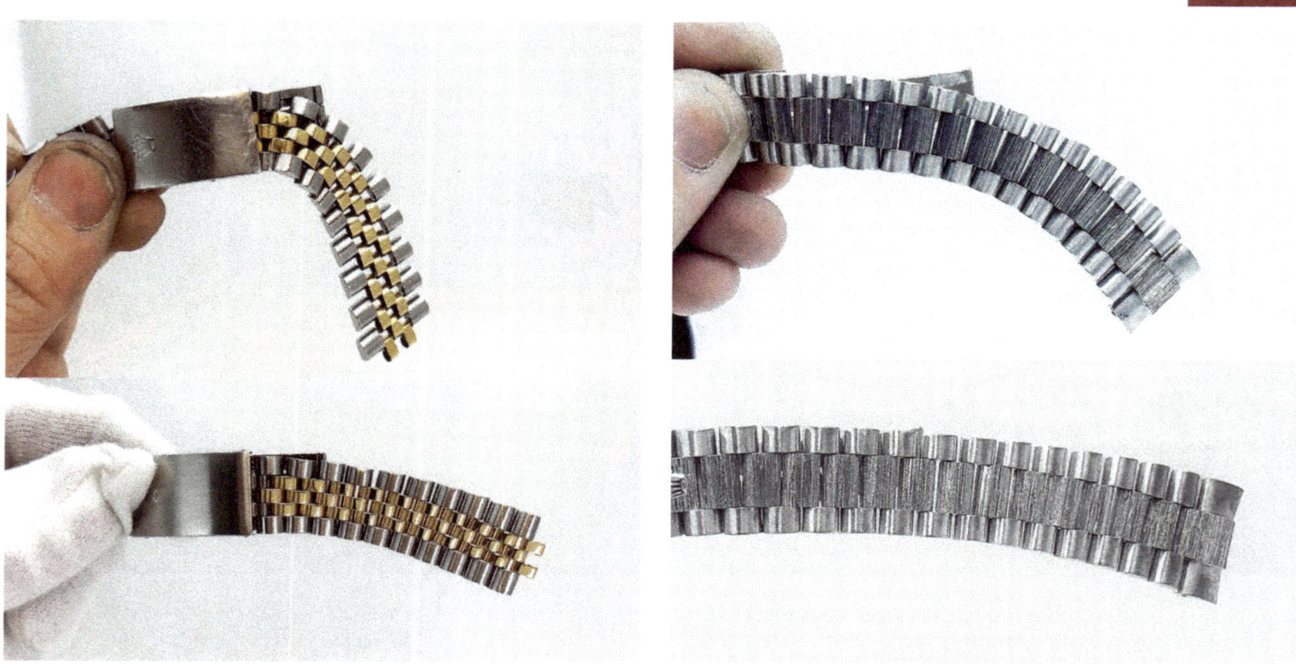

Bracelet stretch (before) and restoration (after). While bracelets are considerd consumable and replaceable parts, they can be restored and reconditioned, complete with exotic finishes (courtesy Rolliworks)

PRESIDENT BRACELET

Rolex introduced the President bracelet on the Day-Date in 1956. It is only available in precious metals and always has a concealed clasp. It is reserved exclusively for the Day-Date and available in different sizes and precious metals.

Variations include the Tridor, which has center links in a mix of three shades of gold. For a brief period in the late 1970s and early 1980s, a distinctive bark finish was available as an option for the center links.

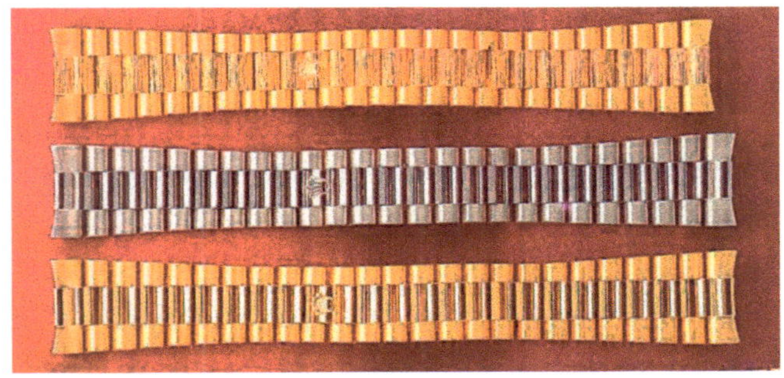

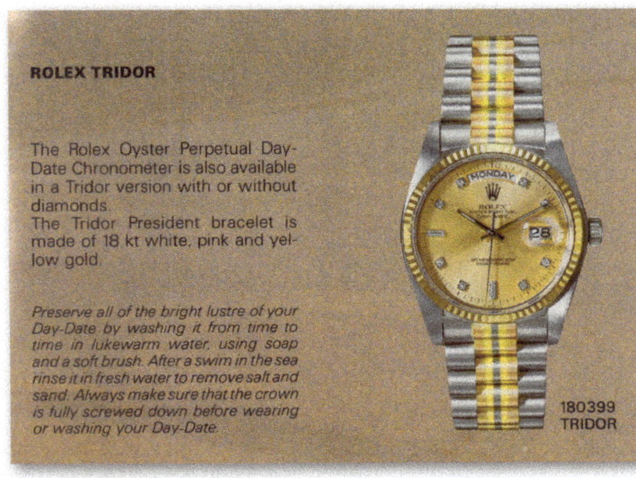

INTEGRATED BRACELETS

These bracelets first appeared on the Oysterquartz and have a distinct angular form, which complements the case features. Like the Cellini Midas, these are claimed to be Gerald Genta designs (though in this case, the claim is controversial and unsupported).

Integrated bracelets are clever and distinctive reinterpretations of the Oyster, Jubilee, and President bracelets. A particularly interesting variation of the President integrated bracelet for the Oysterquartz Day-Date watches involves intricate pyramid patterns.

The steel integrated Oyster bracelet, two-tone integrated Jubilee bracelet, and solid gold integrated President bracelet are close enough to their inspiration to allow them to bear the same names.

PEARLMASTER BRACELET

The Ladies Pearlmaster Collection first appeared in 1992. They are lavish and opulent jewelry bracelets, and often feature jewel settings up to full diamond pave styles to match the cases. They have a rounded five-piece link construction and always have a concealed clasp. They are only available in precious metal.

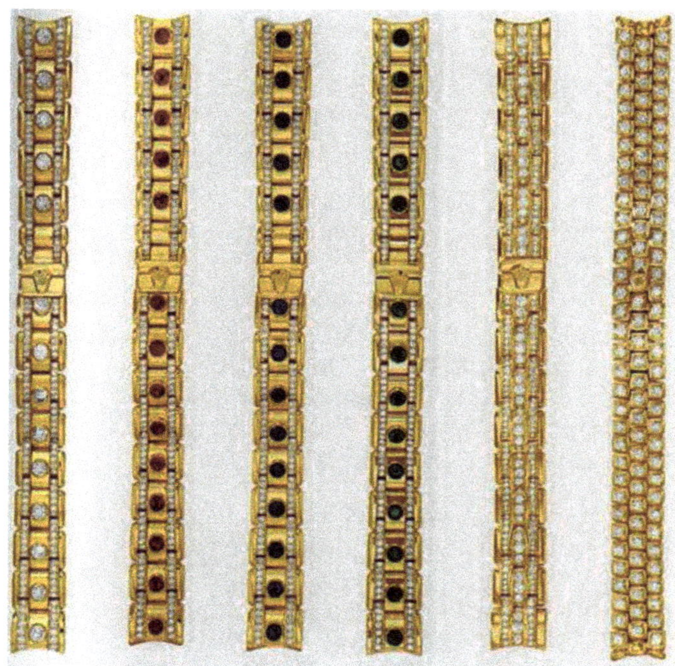

OYSTERFLEX BRACELET STRAP

The Oyserflex is a thoroughly modern strap introduced in 2015 on the Everose Yacht-Master 40mm (ref. 116655) and 37mm (ref. 268655) sizes. It has since been introduced to select Daytona models – ref. 116515LN, the yellow gold Daytona ref. 116518LN, and the white gold Daytona ref. 116519LN.

The Rolex marketing machine insists it is a bracelet rather than a strap, due to the titanium and nickel alloy blade that runs through the middle of the rubber coating. These bracelets (rubber straps) come with the Oysterlock safety clasps, which have a 5mm Easylink extension system. The Oysterflex is replacing leather options on Professional tool watches like the Daytona.

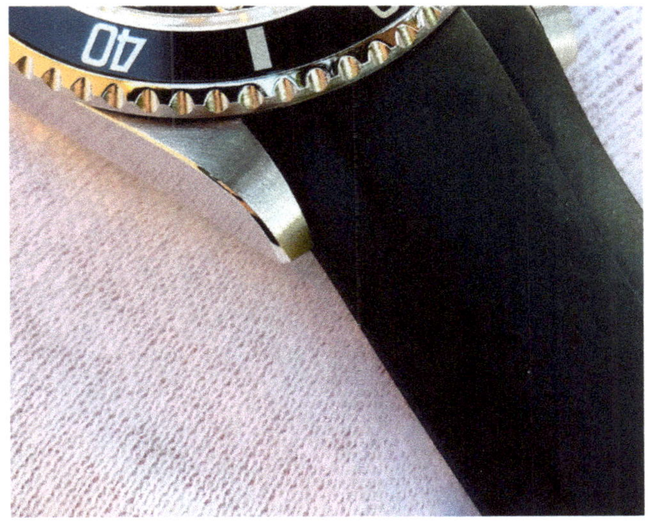

The Oysterflex has yet to make it to other popular Professional models, so manufacturers like Everest Bands are offering credible alternatives. This one features curved ends, eliminating the gap seen on with the Oysterflex.

LEATHER STRAPS

Until the arrival of the Tool watches in the 1950s, leather straps were the norm. Vintage leather straps are seldom worth anything and usually found in unpleasant states of decay.

Modern classics are still available on original leather straps, and in particular, dress watches from the Cellini collection. In the early 2000s, Rolex offered an exclusive white gold Daytona as the Daytona Beach. These featured colorful dials with matching colored straps in pink, turquoise, green, and yellow leather. These Daytona Beach models are rare and quite collectible.

Classic models like the Datejust, Day-Date, and the newer Sky-Dweller can be bought new with leather straps.

END LINKS & SPRING BARS

Rolex introduced end links in 1952 with the launch of the GMT-Master ref. 6542. Their purpose was to reduce movement of the bracelet and pressure on the spring bars, which could result in spring bar failure (and dropping or losing the watch). In addition to improving the streamlined and integrated look of the bracelet, end links increased the durability and reliability of the bracelet. This innovation was necessary for the professional tool watches designed for and used in harsh and demanding environments.

End links are stamped with a number and are specific to the bracelet reference and case size. They also support particular spring bar sizes (diameter). Incorrect, twisted or poorly fitting end links will rattle on the wrist and scratch the Oyster case. Scratching and wear on the case by end links can be bad enough to obscure reference and serial numbers.

Incorrectly-sized spring bars will reduce the effectiveness of end links and contribute to their movement and rattle on the wrist. They can also distort the shape of the lug holes, which can be particularly bad in soft solid gold cases.

End link construction has evolved in step with the bracelets, switching from thin folded steel, to solid milled steel. The solid end link was first used on the Sea-Dweller and has become standard issue.

It is important to use the correct end links and spring bars for the bracelet and case. Ignoring this advice will accelerate wear and tear on both the bracelet and the watch case.

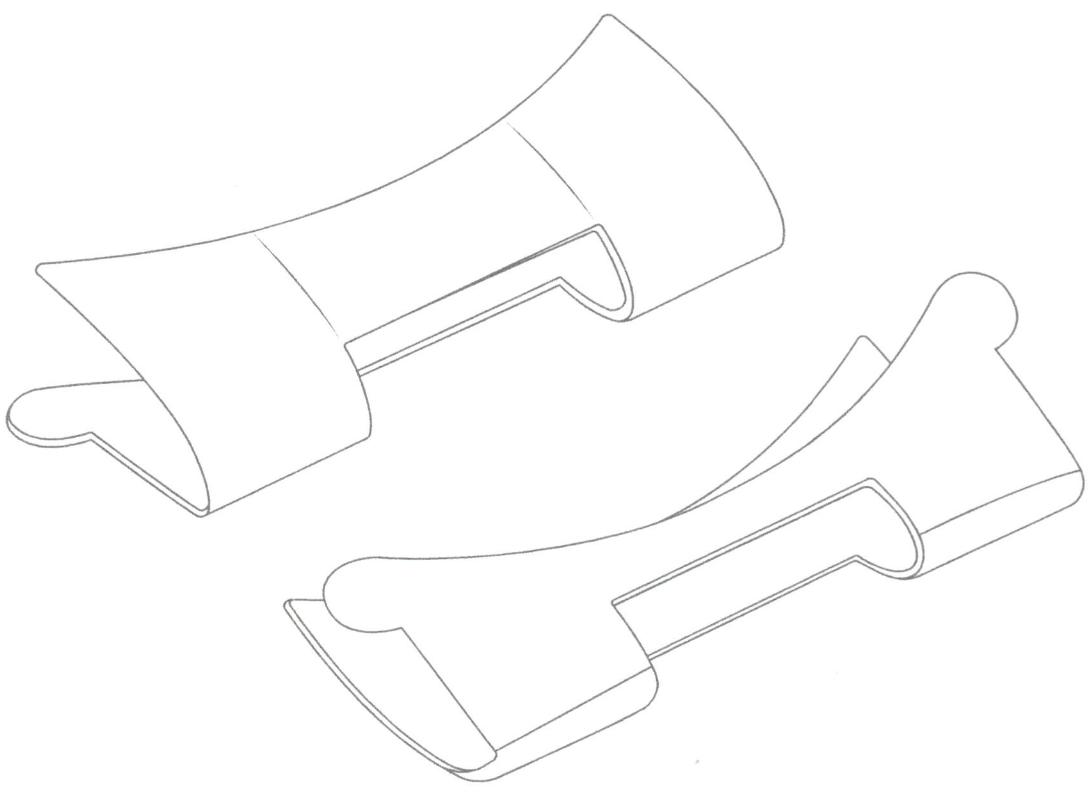

	Part Number	Length	Metal	Model	Pivot Length	Pivot Diameter
	23–9250	11 mm	0	Tudor	1,6 mm	0,9 mm
	23–9260	13 mm	0	—	1,6 mm	0,9 mm
	23–9261	13 mm	1	—	1,6 mm	0,9 mm
	23–9262	13 mm	8	—	1,6 mm	0,9 mm
	23–9263	13 mm	9	—	1,6 mm	0,9 mm
	23–9270	17 mm	0	—	1,7 mm	0,9 mm
	23–9271	17 mm	1	—	1,7 mm	0,9 mm
	23–9272	17 mm	8	—	1,7 mm	0,9 mm
	23–9273	17 mm	9	—	1,7 mm	0,9 mm
	23–9280	19 mm	0	—	1,9 mm	0,9 mm
	23–9281	19 mm	0	—	1,3 mm	0,9 mm
	23–9282	19 mm	8	—	1,9 mm	0,9 mm
	23–9283	19 mm	0	1550/5	0,8 mm	0,9 mm
	23–9290	20 mm	0	—	2,0 mm	0,9 mm
	23–9291	20 mm	0	GMT-Master Submariner	2,8 mm	1,2 mm
	23–9292	20 mm	0 x	GMT-Master	2,8 mm	1,2 mm
	23–9293	20 mm	0	Explorer-1016 Submariner	1,8 mm	1,2 mm
	23–9294	20 mm	8	GMT-Master	2,8 mm	1,2 mm
	23–9350	11 mm	0	Tudor	1,0 mm	0,9 mm
	23–9360	13 mm	0	Tudor	0,9 mm	0,9 mm
	23–9361	13 mm	8	Rolex	1,0 mm	0,9 mm
	23–9362	13 mm	9	Rolex	1,0 mm	0,9 mm
	23–9365	17 mm	8	Rolex	1,0 mm	0,9 mm
	23–9366	17 mm	9	Rolex	1,0 mm	0,9 mm
	23–9370	19 mm	0	—	1,0 mm	0,9 mm
	23–9380	20 mm	0	Day-Date**	1,0 mm	0,9 mm
	23–9381	20 mm	8	Day-Date***	1,2 mm	0,9 mm
	23–9382	20 mm	0	Day-Date***	1,2 mm	0,9 mm
	23–9383	20 mm	9	Day-Date***	1,2 mm	0,9 mm
	23–9384	20 mm	8	Day-Date**	1,0 mm	0,9 mm
	23–9385	20 mm	9	Day-Date**	1,0 mm	0,9 mm
	23–9390	22 mm	Titane	Quartz 5100	1,0 mm	0,9 mm

BRACELET CLASP CODES

From 1976 to 2010, the folding clasp has had a code stamped on the inner blade section, indicating the year of production. Not all vintage bracelets were made in Switzerland, and several subcontractors produced them for specific regions. This practice was mostly for tax purposes and has resulted in a wide variety of code stamps and hallmarks.

Bracelets were not necessarily made the same year as the correct case. Inventory is known to have languished at both the factory and the dealer, so it is acceptable for date stamps to mismatch by a year or two.

The following codes span four decades, covering late vintage and modern classic watches. The rules governing which bracelets could be sold with certain watches became stricter in the 1970s. Before this, retailers and jewelers had considerably more discretion and would routinely swap bracelets and even dials to make a sale. When buying you should check:

that the date code on the clasp is consistent with and within a year or two of the serial number on the watch; and

that the bracelet model was available for sale with that reference.

While these two factors MIGHT indicate the bracelet is original to the watch, it is quite acceptable to wear a new bracelet on an older watch. As long as you know they are mismatched, you can make appropriate price and value adjustments.

Prospective buyers should note that an "S" was also used to denote a Service Replacement clasp. A clasp can be replaced independently of the rest of the bracelet and links.

CLASP DATE CODES (1976–2010)

Year	Code
1976	A or VA
1977	B or VB
1978	C or VC
1979	D or VD
1980	E or VE
1981	F or VF
1982	G
1983	H
1984	I
1985	J
1986	K
1987	L

Year	Code
1988	M
1989	N
1990	O
1991	P
1992	Q
1993	R
1994	S
1995	T or W
1996	V or U
1997	Z or U
1998	Z or W
1999	X

Year	Code
2000	AB
2001	DE
2002	DT
2003	AD
2004	CL
2005	MA
2006	OP
2007	EO
2008	PJ
2009	LT
2010	RS

PROFESSIONAL BRACELETS & END LINKS

The following table will help you know if the bracelet and end links correctly fit your watch. The correct and original combination will depend on the age of the watch, but the following combinations are known to fit.

Some collectors like to wear vintage watches with modern bracelets out of personal comfort preference (solid links and end links). There are no hard and fast rules other than to get a good fit that feels comfortable.

COSMOGRAPH BRACELETS & END LINKS

Case	Bracelet	End Link	Description
6240	6635	71	Stretch Rivet Oyster (19mm)
6240	7205	71	Rivet Oyster (19mm)
6240	7835	271	Folded Oyster (19mm)
6241	6635	71	Stretch Rivet Oyster (20mm)
6241	7205	71	Rivet Oyster (19mm)
6241	7835	271	Folded Oyster (19mm)
6262	7835	71	Folded Oyster (19mm)
6264	7835	271	Folded Oyster (19mm)

DAYTONA BRACELETS & END LINKS

Case	Bracelet	End Link	Description
6239	7205	71	Rivet Oyster (19mm)
6239	7835	771, 271	Folded Oyster (19mm)
6239	78350	771	Solid Oyster (19mm)
6239	6635	71	Stretch Rivet Oyster (19mm)
6239	6251	74	Folded Jubilee (19mm)
6263	7835	771, 271, 371	Folded Oyster (19mm)

Case	Bracelet	End Link	Description
6263	78350	771, 571	Solid Oyster (19mm)
6265	7835	771, 271, 371	Folded Oyster (19mm)
6265	78350	771, 571	Solid Oyster (19mm)
16520	78360	503	Solid Jubilee (20mm)
16520	78390	503B	Solid PCL Oyster (20mm)
16520	78390A	SEL	Solid PCL Oyster (20mm)
116520	78490	SEL	Solid PCL Oyster (20mm)

EXPLORER & EXPLORER II BRACELETS & END LINKS

Case	Bracelet	End Link	Description
1016	7206	58	Rivet Oyster (20mm)
1016	7836	580	Folded Jubilee(20mm)
1016	78360	580	Solid Jubilee (20mm)
14270	78790	558B	Solid Jubilee (20mm)
14270	78690	SEL	Solid Jubilee (20mm)
1655	7206	58	Rivet Oyster (20mm)
1655	7836	580	Folded Oyster(20mm)
1655	78360	580	Solid Oyster (20mm)
16550	7206	58	Rivet Oyster (20mm)
16550	7836	580	Folded Oyster(20mm)
16550	78360	580, 593, 501B	Solid Oyster (20mm)
16550	93150	501B	Solid Oyster (20mm)
16570	78360	501B	Solid Oyster (20mm)
16570	78790	501B	Solid Jubilee (20mm)
16570	78790	SEL	Solid Jubilee (20mm)

GMT-MASTER & GMT-MASTER II BRACELETS & END LINKS

Case	Bracelet	End Link	Description
6542	7206	58, 80	Rivet Oyster (20mm)
6542	6636	58, 64	Stretch Rivet Oyster (20mm)
1675	6636	58, 64	Stretch Rivet Oyster (20mm)
1675	7206	58, 80	Rivet Oyster (20mm)
1675	7836	280	Folded Oyster (20mm)
1675	78360	580	Solid Oyster (20mm)
1675	62510H	550	Solid Jubilee (20mm)
16750	78360	580	Solid Oyster (20mm)
16750	62510H	550	Solid Jubilee (20mm)
16700	78360	501B, 593	Solid Oyster (20mm)
16700	78790A	SEL	Solid Oyster (20mm)
16700	62510H	502B	Solid Jubilee (20mm)
16760	78360	501	Solid Oyster (20mm)
16760	78790A	SEL	Solid Oyster (20mm)
16760	62510H	502B	Solid Jubilee (20mm)
16710	78360	501, 501B	Solid Oyster (20mm)

16710	78790A	SEL	Solid Oyster (20mm)
16710	62510H	502B(T)	Solid Jubilee (20mm)

SEA-DWELLER BRACELETS & END LINKS

Case	Bracelet	End Link	Description
1655	9315	285	Folded Oyster (20mm)
1655	93150	580, 585	Solid Oyster (20mm)
16600	93160	592, SEL	Solid Oyster (20mm)
16660	93160A	SEL	Solid Oyster (20mm)
166600	93160	592, SEL	Solid Oyster (20mm)

SUBMARINER BRACELETS & END LINKS

Case	Bracelet	End Link	Description
5508	7206	58	Rivet Oyster (20mm)
5508	6636	58	Stretch Rivet Oyster (20mm)
6536	7206	80	Rivet Oyster (20mm)
6536	6636	64, 65	Stretch Rivet Oyster (20mm)
6538	6636	64, 65	Stretch Rivet Oyster (20mm)
6538	7206	80	Rivet Oyster (20mm)
6200	6636	64	Stretch Rivet Oyster (20mm)
6200	7206	64	Rivet Oyster (20mm)
6204	6636	64	Stretch Rivet Oyster (20mm)
6204	7206	80	Rivet Oyster (20mm)
6205	6636	64	Stretch Rivet Oyster (20mm)
6205	7206	80	Rivet Oyster (20mm)
5510	6636	80	Stretch Rivet Oyster (20mm)
5510	7206	80	Rivet Oyster (20mm)
5512	7206	80	Rivet Oyster (20mm)
5512	9315	280, 380	Folded Oyster (20mm)
5512	93150	580	Solid Oyster (20mm)
5512	6306	64, 80	Stretch Rivet Oyster (20mm)
5513	7206	80	Rivet Oyster (20mm)
5513	9315	280, 380	Folded Oyster (20mm)
5513	93150	580	Solid Oyster (20mm)
5513	6636	80	Stretch Rivet Oyster (20mm)
1680	9315	280	Folded Oyster (20mm)
1680	93150	580	Solid Oyster (20mm)
1680	7206	80	Rivet Oyster (20mm)
16800	93150	501B	Solid Oyster (20mm)
16800	93250	SEL	Solid Oyster (20mm)
168000	93150	501B	Solid Oyster (20mm)
168000	93250	SEL	Solid Oyster (20mm)
14060/M	93150	501B	Solid Oyster (20mm)
16610	93150	501B	Solid Oyster (20mm)
16610	93250	SEL	Solid Oyster (20mm)

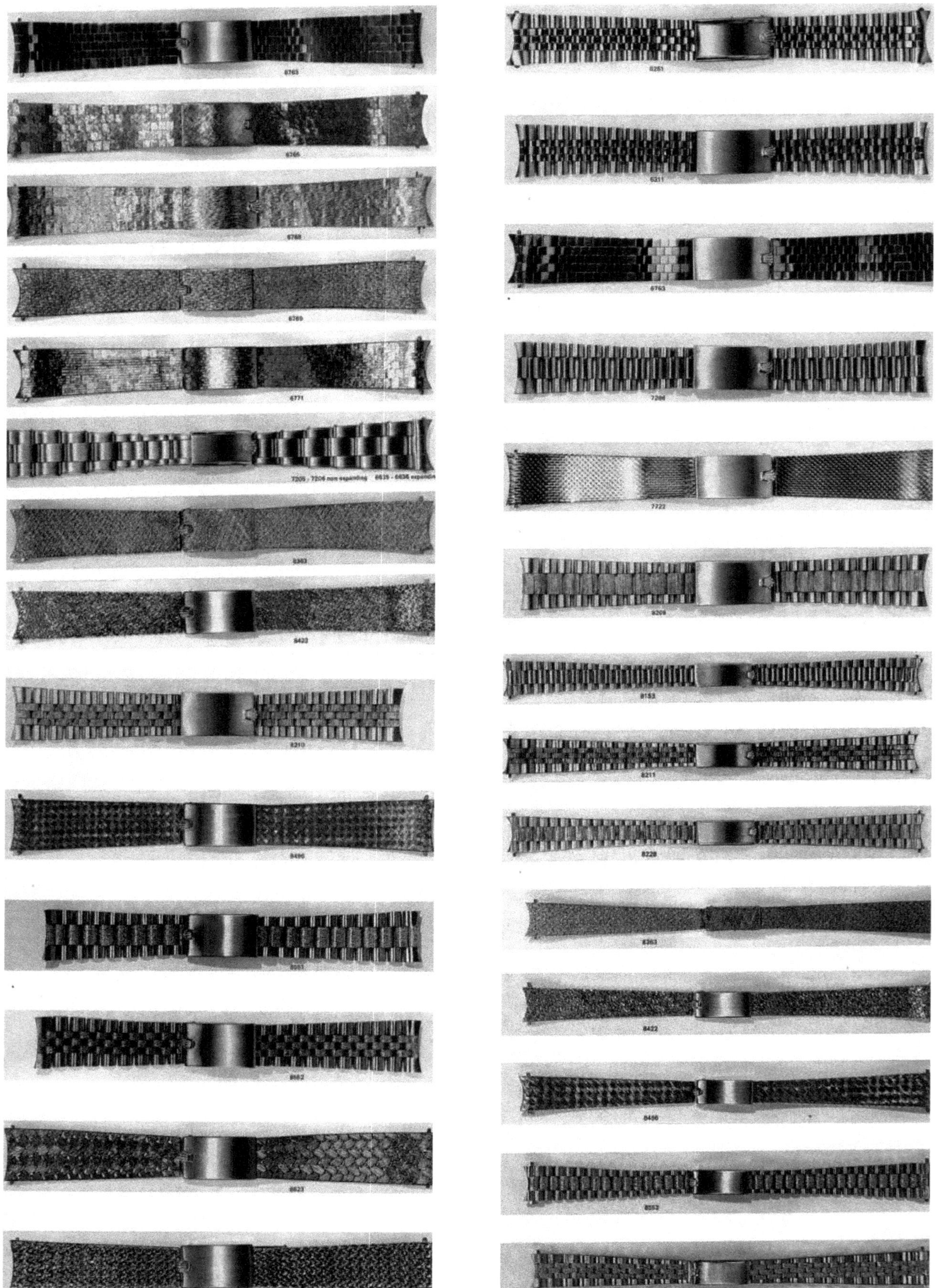

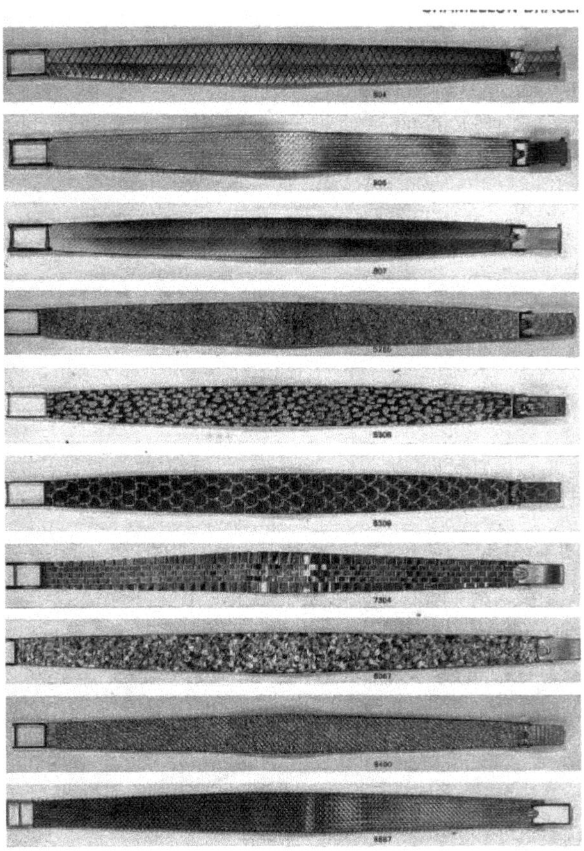

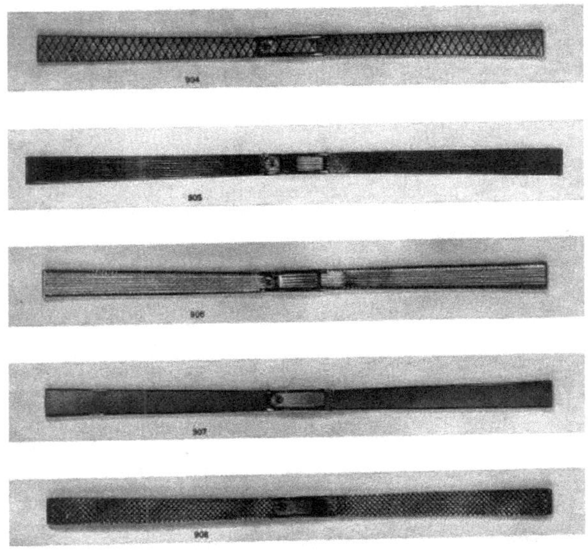

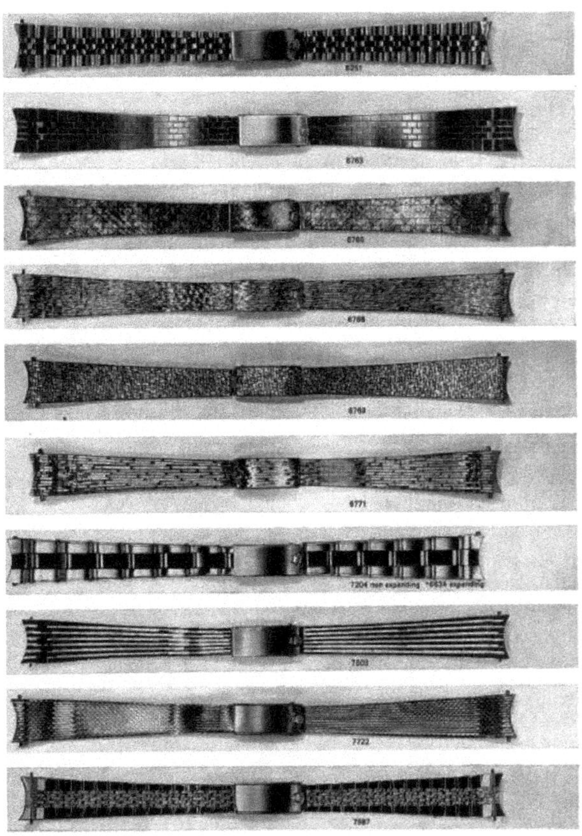

This selection of bracelets and straps is from the 1940s and 1950s. They are typical of the vintage area for four-digit case references.

BRACELET CODES & MODELS

OYSTER-STYLE BRACELETS

Bracelet Ref	Size	Metal	Gender	Links	Description
114	13mm	SS	Ladies		Some riveted
115	13mm	SS	Ladies		Expansion Some riveted
1700/0		SS	Full Size		Oysterquartz Only
201	19mm	SS	Full Size		Straight endsExpansion
202	17mm	SS	Mid Size		Expansion
202	19mm	SS			Some riveted Some expansion
203	20mm	SS	Full Size		Explorer Expansion Some riveted
204	17mm	SS	Mid Size		Straight endsExpansion
206	17mm	SS	Mid Size		Same as 202, but not expansion
206	19mm	SS			Same as 202, but not expansion
207	20mm	SS	Full Size		Explorer, Milgauss, GMT
211	17mm	SS	Mid Size		Expansion
216	17mm	SS	Mid Size		Same as 211, but non-expansion
219	19mm	SS			
301	19mm	SS/14Y			Straight ends: Expansion
302	17mm	SS/14Y	Mid Size		Some riveted Some expansion
302	19mm	SS/14Y			Some riveted Some expansion
306	17mm	SS/14Y	Mid Size		
306	19mm	SS/14Y			Oyster Perpetual Date
308	20mm	SS/14Y	Full Size		
311	17mm	SS/14Y	Mid Size		
314	19mm	SS/14Y			Oyster Perpetual Date
320	13mm	SS114Y	Ladies		
401	19mm	14Pink			
401	19mm	14Y			Some expansion
402	17mm	14Y	Mid Size		Some expansion
402	19mm	14Y			Some riveted Some expansion
405	19mm	14Y			
406	17mm	14Y	Mid Size		Same as 402, but non-expansion
406	19mm	14Y			
411	17mm	14Y	Mid Size		Polished center links
411	19mm	14Y			
420	13mm	14Y	Ladies		Riveted
420	13mm	18Y	Ladies		
501	19mm	10YGF			Straight endsExpansion

Bracelet Ref	Size	Metal	Gender	Links	Description
502	19mm	10YGF			Some riveted Some expansion
506	19mm	10YGF			
514	19mm	10YGF			Some riveted
515	13mm	YGF	Ladies		Tudors
516	13mm	YGF	Ladies		Tudors
6490/8	14mm	18Y	Ladies		100 pink saphires
6490/8	14mm	18Y	Ladies		
6490/9	14mm	18W	Ladies		Riveted Expanding
6490/9	14mm	18W	Ladies		100 pink saphires
6634	13mm	RGP	Ladies		Riveted Expanding
6634/0	11mm	SS	Ladies		Riveted Expanding
6634/0	13mm	SS	Ladies		Riveted Expanding
6634/7	13mm	9Y	Ladies		Riveted Expanding
6634/8	13mm	18Y	Ladies		Riveted Expanding
6635	20mm	RGP	Full Size		Riveted Expanding
6635/0	15mm	SS	Ladies		Riveted Expanding
6635/0	17mm	SS	Mid Size		Riveted Expanding
6635/0	19mm	SS			Riveted Expanding
6635/0	20mm	SS	Full Size		Riveted Expanding
6635/7	20mm	9Y	Full Size		Riveted Expanding
6636/0	20mm	SS	Full Size		Riveted Expanding
7204/7	13mm	14Y	Ladies	13	
7204/8	11 mm		Ladies	13	Some riveted
7204/8	13mm	18Y	Ladies	13	Some riveted
7204/9	13mm	18W	Ladies	13	
7205/7	15mm		Ladies	13	Some riveted
7205/7	17mm	14Y	Mid Size	13	Cosmograph & Oyster Perpetual Date
7205/7	19mm	14Y		13	Some riveted
7205/8	17mm	18Y	Mid Size	13	Cosmograph & Oyster Perpetual Date
7205/8	19mm	18Y		13	Cosmograph
7205/8	20mm	18Y	Full Size	13	Some riveted
7205/9	17mm	18W	Mid Size	13	Some riveted
7205/9	19mm	18W		13	Some riveted
7205/9	20mm	18W	Full Size	13	Some riveted
7206/7	20mm	14 Y	Full Size	13	Datejust & GMT II
7206/8	20mm	18Y	Full Size		
7215/5	19mm	YGF			
7295/8 Bic	14mm	Tridor	Ladies		

OYSTER-STYLE BRACELETS CONT.

Bracelet Ref	Size	Metal	Gender	Links	Description
7295/9	14mm	18W	Ladies		
7385/5	20mm	18R	Full Size		President concealed clasp
7385/8	20mm	18Y	Full Size		President concealed clasp
7497/8	14mm	18Y	Ladies		270 diamonds
7497/9	14mm	18W	Ladies		270 diamonds
7498/8	14mm	18Y	Ladies		286 diamonds
7498/9	14mm	18W	Ladies		286 diamonds
7805/0	13mm	SS	Ladies		
7805/0	17mm	SS	Mid Size		
7805/3	13mm	SS/18Y	Ladies		
7805/3	17mm	SS/18Y	Mid Size		
7824/0	13mm	SS	Ladies		
7824/3	13mm	SS/18Y	Ladies		
7834/0	11mm	SS	Ladies	13	
7834/0	13mm	SS	Ladies	13	
7834/1	11mm	YGF	Ladies	13	
7834/1	13mm	YGF	Ladies	13	
7834/3	11mm	SS/14Y	Ladies	13	
7834/3	13mm	SS/14Pink	Ladies	13	
7834/3	13mm	SS/14Y	Ladies	13	
7834/3	13mm	SS/18Y	Ladies	13	
7835/0	17mm	SS	Mid Size	13	
7835/0	19mm	SS		13	Cosmograph, Air King & Oyster Perpetual
7835/1	19mm	YGF		13	
7835/3	17mm	SS/14Y	Mid Size	13	
7835/3	17mm	SS/18Y	Mid Size	13	
7835/3	19mm	SS/14Pink		13	
7835/3	19mm	SS/14Y		13	
7835/3	19mm	SS/18Y		13	Oyster Perpetual
7836/0	20mm	SS	Full Size	13	Milgauss, Datejust GMT, Explorer
7836/1	20mm	YGF	Full Size	13	Datejust
7836/3	20mm	SS/14Y	Full Size	13	GMT & Datejust
7836/3	20mm	SS/18Y	Full Size	13	GMT & Datejust
7866/8	20mm	18Y	Full Size	13	Daytona
7879/0	20mm	SS	Full Size		Explorer II, GMT & GMT II
7879/3	20mm	SS/18Y	Full Size		GMT II
7879/8	20mm	18Y	Full Size		GMT II
8363/8	13mm	18Y	Ladies		Moire finish

Bracelet Ref	Size	Metal	Gender	Links	Description
8606/8	13mm	18Y	Ladies		
9290/8	20mm	18Y	Full Size	12	GMT & Submnariner Fliplock
9315/0	20mm	SS	Full Size	12	GMT & Submnariner Fliplock
9315/3	20mm	SS/18Y	Full Size	12	Submariner Fliplock
9316/0	20mm	SS	Full Size	14	Sea-Dweller Fliplock
9316A/O	20mm	SS	Full Size	14	Sea-Dweller Fliplock
9325/0	20mm	SS	Full Size		Explorer & Submariner Fliplock
9325/3	20mm	SS/18Y	Full Size		Submariner Fliplock
9351/0	20mm	SS	Full Size		Submariner Fliplock
7876/8	20mm	18Y	Full Size		Yacht-Master Fliplock
7839/0	20mm	SS	Full Size		Daytona Fliplock
7839/3	20mm	SS/18Y	Full Size		Daytona Fliplock
7839/3A	20mm	SS/18Y	Full Size		Daytona Solid end links (SEL)
7839/8	20mm	18Y	Full Size		Daytona Fliplock
7849/0	20mm	SS	Full Size		Daytona Solid end links (SEL)
7873/0	14mm	SS	Ladies		Yacht-Master Only
7873/3	14mm	SS/18Y	Ladies		Yacht-Master Only
7873/8	14mm	18Y	Ladies		Yacht-Master Only
7874/0	17mm	SS	Mid Size		Yacht-Master
7874/3	17mm	SS/18Y	Mid Size		Yacht-Master
7874/8	17mm	18Y	Mid Size		Yacht-Master
7875/0	17mm	SS	Mid Size		Yacht-Master
7875/3	17mm	SS/18Y	Mid Size		Yacht-Master
7875/8	17mm	18Y	Mid Size		Yacht-Master
7876/0	20mm	SS	Full Size		Yacht-Master Fliplock
7879/0	20mm	SS	Full Size		Explorer
7879/3	20mm	SS/18Y	Full Size		Explorer
7879/8	20mm	18Y	Full Size		
7879/OA	20mm	SS	Full Size		Daytona Solid end links (SEL)

JUBILEE-STYLE BRACELETS

Style	Bracelet Ref	Size	Metal	Gender	Links	Description
Jubiiee	**6251/3**	17mm	SS/14K	Mid Size		
Jubiiee	**6251/3**	19mm	SS/14Pink			
Jubilee	**109**	20mm	18YGF	Full Size		Datejust & Thunderbird
Jubilee	**1701/0**		SS	Full Size	13	Oysterquartz Only
Jubilee	**1701/3**		SS/18Y	Full Size	13	Oysterquartz Only
Jubilee	**200**	20mm	SS	Full Size		Folded links
Jubilee	**200**	20mm	SS/YG	Full Size		Folded links
Jubilee	**201**	13mm	SS	Ladies		Folded links
Jubilee	**208**	20mm	SS	Full Size		Datejust
Jubilee	**209**	19mm	SS			
Jubilee	**210**	13mm	SS	Ladies		
Jubilee	**218**	20mm	SS	Full Size		Datejust
Jubilee	**219**	17mm	SS	Mid Size		
Jubilee	**303**	20mm	SS/14Y	Full Size		Datejust & Thunderbird
Jubilee	**304**	19mm	SS/14Y			
Jubilee	**308**	20mm	SS/14Y	Full Size		Datejust & GMT
Jubilee	**310**	13mm	14Y	Ladies		
Jubilee	**310**	13mm	SS/14Y	Ladies		
Jubilee	**313**	20mm	SS/14Y	Full Size		Datejust & GMT
Jubilee	**314**	17mm	SS/14Y	Mid Size		
Jubilee	**314**	19mm	SS/14Y	Mid Size		
Jubilee	**403**	19mm	14Y	Mid Size		
Jubilee	**405**	17mm	14Y	Mid Size		Oyster Perpetual Date
Jubilee	**405**	19mm	14Y			
Jubilee	**410**	13mm	14 Y	Ladies		
Jubilee	**410**	13mm	14W	Ladies		
Jubilee	**6251/0**	13mm	SS	Ladies		
Jubilee	**6251/0**	17mm	SS	Mid Size	20	
Jubilee	**6251/0**	19mm	SS		22	Oyster Perpetual Date
Jubilee	**6251/0**	20mm	SS	Full Size	22	GMT & Datejust
Jubilee	**6251/3**	13mm	SS/14K	Ladies		
Jubilee	**6251/3**	13mm	SS/14Pink	Ladies		
Jubilee	**6251/3**	13mm	SS/18K	Ladies		
Jubilee	**6251/3**	19mm	SS/14K			
Jubilee	**6251/3**	19mm	SS/18K			
Jubilee	**6251/3**	20mm	SS/14K	Full Size		

Style	Bracelet Ref	Size	Metal	Gender	Links	Description
Jubilee	**6251/3**	20mm	SS/14Pink	Full Size		
Jubilee	**6251/3**	20mm	SS/18K	Full Size		
Jubilee	**6251/7**	13mm	14Y	Ladies	21	
Jubilee	**6251/8**	13mm	18Y	Ladies	21	
Jubilee	**6251/9**	13mm	18W	Ladies	21	
Jubilee	**6252/3**	13mm	SS/14Pink	Ladies		
Jubilee	**6252/3**	13mm	SS/14K	Ladies		
Jubilee	**6252/3**	13mm	SS/18K	Ladies		
Jubilee	**6252/3**	17mm	SS/18K	Mid Size		
Jubilee	**6252/3**	19mm	SS/14K			
Jubilee	**6252/3**	19mm	SS/18K			
Jubilee	**6252/3**	20mm	SS/14Pink	Full Size		
Jubilee	**6252/3**	20mm	SS/14Y	Full Size		GMT
Jubilee	**6252/3**	20mm	SS/18Y	Full Size		GMT & Datejust
Jubilee	**6311/0**	17mm	SS	Mid Size		
Jubilee	**6311/0**	20mm	SS	Full Size	23	
Jubilee	**6311/3**	17mm	SS/18Y	Mid Size		
Jubilee	**6311/3**	20mm	SS/18Y	Full Size		
Jubilee	**6311/7**	20mm	14Y	Full Size		
Jubilee	**6311/8**	17mm	18Y	Mid Size	26	
Jubilee	**6311/8**	19mm	18Y		21	
Jubilee	**6311/8**	20mm	18Y	Full Size	21	GMT, Oyster Perpetual, Datejust & Submariner
Jubilee	**6311/9**	17mm	18W	Mid Size		
Jubilee	**6311/9**	20mm	18Y	Full Size		Datejust
Jubilee	**8210/7**	20mm	14Y	Full Size	21	Engraved
Jubilee	**8210/8**	20mm	18Y	Full Size	21	Bark finish
Jubilee	**8211/7**	13mm	14Y	Ladies	21	Bark finish
Jubilee	**8211/8**	13mm	18Y	Ladies		Bark finish
Jubilee	**8211/9**	13mm	18W	Ladies		Bark finish
Jubilee	**8386/7**	20mm	14Y	Full Size		Concealed clasp
Jubilee	**8386/8**	20mm	18Y	Full Size	24	Concealed clasp Datejust
Jubilee	**8387/8**	20mm	18Y	Full Size		Concealed clasp GMT & Submariner
Jubilee	**8391/8**	17mm	18Y	Mid Size	28	Concealed clasp
Jubilee	**8552/8**	20mm	18Y	Full Size		Bark finish
Jubilee	**8554/8**	13mm	18Y	Ladies	21	Moire finish
Jubilee	**8554/9**	13mm	18W	Ladies		Engraved

JUBILEE-STYLE BRACELETS CONT.

Style	Bracelet Ref	Size	Metal	Gender	Links	Description
Jubilee	**8571/8**	13mm	18Y	Ladies	36	Concealed clasp
Jubilee	**9667/3**		SS/14Y	Full Size	13	Oysterquartz Only
Super Jubilee	**6411/8**	17mm	18Y	Mid Size	28	263 diamonds in center links
Super Jubilee	**6411/9**	17mm	18W	Mid Size	28	263 diamonds in center links
Super Jubilee	**6411/9 Bic**	17mm	Tridor	Mid Size	28	263 diamonds in center links
Super Jubilee	**6451/8**	13mm	18Y	Ladies	36	344 diamonds in center links
Super Jubilee	**6451/9**	13mm	18W	Ladies	36	344 diamonds in center links:
Super Jubilee	**6451/9 Bic**	13mm	Tridor	Ladies		
Super Jubilee	**6453/9**	13mm	18W	Ladies	38	360 diamonds
Super Jubilee	**6454/8**	13mm	18Y	Ladies		89 diamonds
Super Jubilee	**6454/9**	13mm	18W	Ladies		89 diamonds
Super Jubilee	**8486/8**	20mm	18Y	Full Size	23	218 diamonds in center links
Super Jubilee	**8486/9**	20mm	18W	Full Size	23	218 diamonds in center links
Super Jubilee	**8486/9 Bic**	20mm	Tridor	Full Size	23	218 diamonds in center links
Super Jubilee Karat	**6453/8**	13mm	18Y	Ladies	38	360 diamonds

PRESIDENT-STYLE BRACELETS

Style	Bracelet Ref	Size	Metal	Gender	Links	Description
President	**1901/8**		18Y	Full Size	25	Oysterquartz Day-Date Only
President	**1901/9**		18W	Full Size	25	Oysterquartz Day-Date Only
President	**1902/8**		18Y	Full Size	25	Oysterquartz Day-Date Only
President	**1914/8**		18Y	Full Size	25	Oysterquartz Day-Date Only
President	**7274/6**	20mm	Platinum	Full Size		Concealed clasp
President	**7274/8**	20mm	18Y	Full Size		Concealed clasp
President	**7286/8**	20mm	18Y	Full Size	21	Regular clasp
President	**7286/9**	20mm	18W	Full Size	21	Regular clasp
President	**8153/8**	13mm	18Y	Ladies	32	
President	**8153/9**	13mm	18W	Ladies		
President	**8209/8**	20mm	18Y	Full Size	21	Bark finish Regular clasp
President	**8209/9**	20mm	18W	Full Size	21	Bark finish Regular clasp
President	**8211/8**	13mm	18Y	Ladies		Bark finish
President	**8211/9**	13mm	18W	Ladies		Bark finish
President	**8228/8**	13mm	18Y	Ladies	31	Bark finish
President	**8228/9**	13mm	18W	Ladies	31	Bark finish
President	**8270/9**	13mm	Tridor	Ladies	36	
President	**8285/9**	20mm	Tridor	Full Size	24	Concealed clasp

Style	Bracelet Ref	Size	Metal	Gender	Links	Description
President	**8289/8**	20mm	18Y	Full Size		Concealed clasp
President	**8289/9**	17mm	Tridor	Mid Size	28	Concealed clasp
President	**8385/6**	20mm	Platinum	Full Size	24	Concealed clasp
President	**8385/8**	20mm	18Y	Full Size	24	Concealed clasp
President	**8385/8**	20mm	18Y	Full Size	24	Concealed clasp Baguette diamonds
President	**8385/9**	20mm	18W	Full Size	24	Concealed clasp
President	**8385/9**	20mm	18W	Full Size	24	Concealed clasp Baguette diamonds
President	**8389/6**	17mm	Platinum	Mid Size	28	Concealed clasp
President	**8389/8**	17mm	18Y	Mid Size	28	Concealed clasp
President	**8389/9**	17mm	18W	Mid Size	28	Concealed clasp
President	**8390/6**	17mm	Platinum	Mid Size	24	Concealed clasp
President	**8390/8**	17mm	18Y	Mid Size		Concealed clasp
President	**8390/9**	17mm	18W	Mid Size		Concealed clasp
President	**8553/8**	13mm	18Y	Ladies	31	Engraved
President	**8570/6**	13mm	Platinum	Ladies	36	Concealed clasp
President	**8570/8**	13mm	18Y	Ladies	36	Concealed clasp
President	**8570/8**	13mm	18Y	Ladies		Concealed clasp Baguette diamonds
President	**8570/9**	13mm	18W	Ladies	36	Concealed clasp
President	**8723/8**	20mm	18Y	Full Size	24	Concealed clasp Bark finish
President	**8723/9**	20mm	18W	Full Size	24	Concealed clasp Bark finish
President	**9235/8**	13mm	18Y	Ladies	36	Concealed clasp Bark finish
President	**9235/9**	13mm	18W	Ladies	36	Concealed clasp Bark finish
President Karat	**8472/8**	13mm	18Y	Ladies		Diamond stripes
Super President	**8470/6**	13mm	Platinum	Ladies	36	Concealed clasp Diamond center links
Super President	**8470/8**	13mm	18Y	Ladies		Concealed clasp Diamond center links
Super President	**8470/9**	13mm	18W	Ladies		Concealed clasp Diamond center links
Super President	**8485/6**	20mm	Platinum	Full Size	23	288 diamonds in center links
Super President	**8485/8**	20mm	18Y	Full Size	23	288 diamonds in center links
Super President	**8485/9**	20mm	18W	Full Size	23	288 diamonds in center links
Super President	**8489/6**	17mm	Platinum	Mid Size	28	340 diamonds in center links
Super President	**8489/8**	17mm	18Y	Mid Size	28	340 diamonds in center links
Super President	**8489/9**	17mm	18W	Mid Size	28	340 diamonds in center links
Super President Karat	**8473/8**	13mm	18Y	Ladies		446 diamonds

OTHER BRACELET STYLES

Style	Bracelet Ref	Size	Metal	Gender	Links	Description
	101/8	17mm	18Y	Mid Size		
	101/9	17mm	18W	Mid Size		
	102/8	17mm	18Y	Mid Size		
Bark	**100/8**	13mm	18Y	Ladies		
Bark	**100/9**	13mm	18W	Ladies		
Basle Fair	**704**		14Y	Ladies		Oyster Perpetual & Ladydate
Brick	**705**		14Y	Ladies		
Brick	**706**		14W	Ladies		
Chameleon	**705**		14W	Ladies		Chameleon
Chameleon	**706**		14Y	Ladies		Chameleon
Five-link	**2563/7**	20mm	9Y	Full Size		Capture clasp
Flat Squares	**2557/7**	20mm	9Y	Full Size		Capture clasp
Ladies	**101/8**	13mm	18Y	Ladies		
Ladies	**102/8**	13mm	18W	Ladies		
Ladies	**102/9**	13mm	18W	Ladies		
Ladies	**114**	13mm	SS	Ladies		
Matte weave	**2558/7**	20mm	9Y	Full Size		Capture clasp
Mesh	**102/8**	19mm	18Y			Moire finish
Mesh	**704**		14Y	Ladies		
Mesh	**715**	20mm	18Y	Full Size		
Mesh	**716**	19mm	18Y			
Mesh	**71770**	13mm	14Y	Ladies		Chameleon
Mesh	**71779**	13mm	14W	Ladies		Chameleon
Mesh	**725**	19mm	14Y			
Mesh	**732**		14Y	Ladies		
Mesh	**7435/8**		18Y			Flushfit for Non-oyster models
Mesh	**7435/9**		1W			Flushfit for Non-oyster models
Mesh	**7491**		18W	Mens		
Mesh	**7491**		18Y	Mens		
Mesh	**7491/8**		18W			Fits only case ref# 9578 dress watch
Mesh	**7491/9**		18W			Fits only case ref# 9578 dress watch
Mesh	**8363/7**	13mm	14Y	Ladies		Brushed mesh for Datejust
Mesh	**8363/7**	17mm	14Y	Mid Size		
Mesh	**8363/7**	17mm	18Y	Mid Size		
Mesh	**8363/7**	19mm	14 Y			

Style	Bracelet Ref	Size	Metal	Gender	Links	Description
Mesh	**8363/8**	13mm	18Y	Ladies		
Mesh	**8363/8**	13mm	18Y	Ladies		
Mesh	**8606/8**	13m	18Y	Ladies		Loose mesh
Mesh	**8607/8**	20mm	18Y	Full Size		Loose mesh
Mesh	**R24**	17mm	14Y	Mid Size		
Tri-link	**2544/7**	20mm	9Y	Full Size		Capture clasp
Tri-link	**6490/8**	13mm	18Y	Ladies		
Yacht-Master	**7294/8**	14mm	18Y	Ladies		
Yacht-Master	**7294/8**	14mm	Tridor	Ladies		
Yacht-Master	**7294/9**	14mm	18W	Ladies		
Yacht-Master	**7490/8**	14mm	18Y	Ladies		
Yacht-Master	**7490/8**	14mm	18Y	Ladies		Center Emerald Diamond outer row
Yacht-Master	**7490/8**	14mm	18Y	Ladies		Center Ruby Diamond outer row
Yacht-Master	**7490/8**	14mm	18Y	Ladies		Center Sapphire Diamond out row
Yacht-Master	**7490/9**	14mm	18W	Ladies		Diamonds
Yacht-Master	**7494/8**	14mm	18Y	Ladies		Diamond outer row
Yacht-Master	**7494/9**	14mm	18W	Ladies		Diamond outer row
Yacht-Master	**7495/8**	14mm	Tridor	Ladies		15 diamonds

Jubilee | jü-bə-ˈlē |

noun ***historical***

- a season of celebration
- a celebration of a special anniversary
- an iconic Rolex ***bracelet***

OYSTER WATCH Co
GENEVA-SWISS
GRANT BRITAIN PATENTS
SWISS PATENTS
FRENCH PATENTS
U.S.A PATENTS
GERMAN PATENTS

1932 - 1945

Chapter 11

MOVEMENTS & CALIBERS

Most collectors don't know or care much about the movement in their beautiful vintage heirloom, and providing that it keeps good time, there's little need to worry. However, unscrupulous practices can rob an owner or buyer of significant value so it helps to have some sense of what the inner parts of the watch should be and how to identify them.

Experienced collectors say that before paying for a vintage watch, it is essential to inspect what lies beneath the caseback visually – or at the very least, see high-quality photos of the watch's inner workings. It is common practice to ask for (and get) high-quality movement photos during purchase negotiations and even during service. Any refusal should be considered suspicious.

Dealers will often claim that opening the watch will break the integrity of the waterproof seal or that they don't have the tool to open it. While this may be true, it is also a warning that they might be trying to conceal something. In these instances, confirm their return policy in writing and make it clear you plan to have the piece opened and inspected as soon as you receive it.

There is some risk in allowing an independent watchmaker to work on your vintage watch. It is not difficult or uncommon to substitute generic, aftermarket, or even fake components for rare and valuable vintage parts. Watchmakers who have built a career and reputation for working on vintage Rolex should be sought out by engaging the collector community and asking for recommendations and referrals. Don't send your precious Rolex heirloom to just any old watchmaker who says they can work on a Rolex. Whether they are just inspecting, or servicing the watch, they will need to be equipped with specialized tools, vintage Rolex service manuals, and have access to tightly restricted parts.

Ideally, your watchmaker should be known to the collector community, professionally certified or accredited, and willing to communicate openly and often throughout the service process. This communication typically includes progress photos of your watch in various states of disassembly

A trusted and recognized vintage Rolex watchmaker nearly always has a backlog of work and impatient clients. The technical work typically takes a few days,

"A body in motion can maintain this motion only if it remains in contact with a mover"
Aristotle

and they will have several concurrent projects on their bench in various states of assembly. Progress can halt if they are waiting on parts, and it can take several months to turn your watch around. Some movements are more challenging and can take longer than others.

Each movement is uniquely numbered (like cases), but it is not currently possible to match a movement number with a case serial number. Rolex has never released this data, so it is impossible to say whether a movement left the factory in the watch case it currently inhabits. However, an experienced vintage Rolex watchmaker will be able to determine if the movement number falls within the expected range for the case number.

For antique movements with no movement number, a watchmaker should be able to assess the finish of the movement and say if it's consistent for the manufacturer, the reference, and the era. This determination is subjective, technical, and requires a lot of experience. Like any opinion, it is open to controversy and debate.

Availability of antique and vintage movement parts is a growing challenge. Collectors and watchmakers often hoard new genuine (NOS) vintage parts as well as donor movements. Donor movements and parts will occasionally show up on eBay, and this is not an uncommon source of parts for a watchmaker. So, offering to help your watchmaker locate parts can reduce the turnaround time of your watch, allowing them to focus on the work rather than eBay. Sending along a donor watch or parts can help, too.

Contemporary modern classics have better production records, and an RSC can check if a movement number matches the case number. Unless the movement is entirely Franken (or fake), they are likely to replace any aftermarket parts and bring it up to their required quality standards.

A watchmaker can order new parts from Rolex if they have access to a coveted Parts Account (or a fellow watchmaker with said access). Rolex has increasingly strict rules governing who can buy parts, and independent watchmakers can struggle to get what your watch needs. Before sending your vintage watch for service, ask the watchmaker how they source their parts.

Rolex relied on parts from numerous subcontractors and suppliers over the years – up to 23 distinct entities at one point. Movements came predominantly from Aegler, except for chronograph movements supplied by Valjoux and later Zenith. These are ébauche, or generic base movements, and were intended to be customized and finished by the watchmakers.

Today, Rolex makes everything at only three factories in Switzerland—Bienne for the movements, Plan-Les-Ouates for cases and bracelets, and Les Acacias for final assembly and testing. This fully-integrated supply chain is a modern phenomenon and has only been in existence since 2004.

Rolex has been buying movements from its close partner, Aegler, since Rolex's inception in 1908, and finally acquired the Aegler company in 2004. It wasn't until this merger with their most influential supply chain partner that Rolex was able to produce genuine in-house movements. So for vintage and antique Rolex watches, movements are third-party (mostly Aegler) with parts branded and finished by, or for, Rolex.

In the past, Aegler was not an exclusive supplier to Rolex, and there were periods (pre-1950) in which they could not meet demand and Rolex had to source movements and parts from other manufacturers like Beguelin. Because of this, it should be pointed out that even though an antique Rolex has a non-Aegler movement does not mean it is fake or a Frankenwatch.

Rolex deployed Aegler movements in a wide variety of references spanning half a century (antique, vintage, and modern classic era). While it makes for a convoluted family tree, it adds interest and variety to a challenging combination of possible configurations.

THE SERVICE & RESTORATION BUSINESS

The craft of watchmaking is rooted in engineering practices. As movement complications grow more sophisticated, the craft strays into the realms of fine art. This is borne out in incredible over-engineering and finishing so characteristic of other Swiss watchmakers. The pursuit of this art and the quest to push boundaries is known as "haute horology". Rolex watches are not considered haute horology and the watch cognoscenti describe them as industrial, utilitarian, and unrefined. This does not diminish the enthusiasm of vintage Rolex collectors or the pre-owned market. These watches are desired as much for their history as their craftsmanship.

The businesses of today's young independent craftsmen and women, are diverse and specialized. Some, focus on movements. Others, specialize in restoration and finishing. Still others, pursue sub-specialties like dial refinishing and reluming.

When seeking an independent craftsman to work on your vintage Rolex, consider their reputations for customer satisfaction and work quality. Their rates and fees are generally not negotiable. As with all things Rolex, seeking to save money and cut corners can result in erosion of the value of your vintage watch and ultimately disappointment. It should be noted that independent professionals are not necessarily cheaper, nor do they perform lesser quality work than their official and authorized counterparts. They're just not regulated, controlled, or otherwise affiliated with Rolex.

While it may be tempting or even obvious to take your vintage Rolex watch to a Rolex Service Center, this is contrary to popular collector opinion. The oldest movement they will service (at the time of writing) is the caliber series 15xx. With diminishing parts this is unlikely to last. They are also known to insist on replacing cosmetically distressed parts (hands, dials, crowns, bezel inserts) that are often most prized by vintage enthusiasts.

For high-value vintage pieces, Rolex Geneva offers a restoration service. An estimate for work will cost approximately $1,500 (US) and take up to 6 months to perform. If the work is executed this fee is waived and the restoration project can take up to two years. Where parts are unavailable, Rolex will custom make them, retaining the claim to being Swiss Made. With the high price and long turn around time, this service is not for the casual collector but for historically important pieces.

For Modern Classic and contemporary movements, the collectors options are broader. In addition to the independent professionals, Official Rolex Jewelers (ORJs) can perform servicing to the same standards as Official Rolex Service Centers and issue the same service warranty documents and guarantees.

An ORJ must have a Rolex trained and certified watchmaker on staff to qualify for the Plaque Program (literally a plaque on a jewelers store front). An ORJ can qualify as a Rolex Plaque holder at one of four levels.

Level 30 is most common and certifies an ORJ to service all contemporary Oyster case models.

Level 40 includes all models from the Professional Collection and requires the ORJ to own and operate lapping equipment for case finishing.

Level 50 certifies the ORJ to service chronograph movements – caliber 4030 and 4131 used in modern Daytonas.

Level 60 permits servicing of the Yacht-Master II and the Skydweller.

LUBRICANTS

Movements are powered by the unwinding of a tightly-coiled flat spring that drives a train of gears. Unless they are left to run down, the parts are under constant pressure and in continuous motion. Several types of lubricants are used to reduce wear and maximize the longevity of the movement. The specific type, quantity, and location of the oils are specified in the watchmaker's lubrication schedule, and these old service manuals are becoming as sought after as the watches themselves.

A movement service of a caliber 3035 and 3135 calls for up to six different lubricants (!), ranging from light oil to grease.

Lubricants will break down over time, irrespective of whether the movement is running. Modern synthetic oils last longer but experience the same gradual degradation. With the passing of time and an inactive movement, lubricants can move and pool (thanks to gravity), so it's considered good practice to wind a vintage movement every few weeks.

Desiccated lubricants can be difficult to clean from small parts, with remaining traces causing unwanted friction and final tuning problems. Watchmakers go to great lengths with sophisticated cleaning machines and harsh cleaning agents to remove old lubricants. This often tricky task is one of the arguments in favor of regular servicing. Current

A MODERN INDY CW21

Phillip Ridley, TX, USA

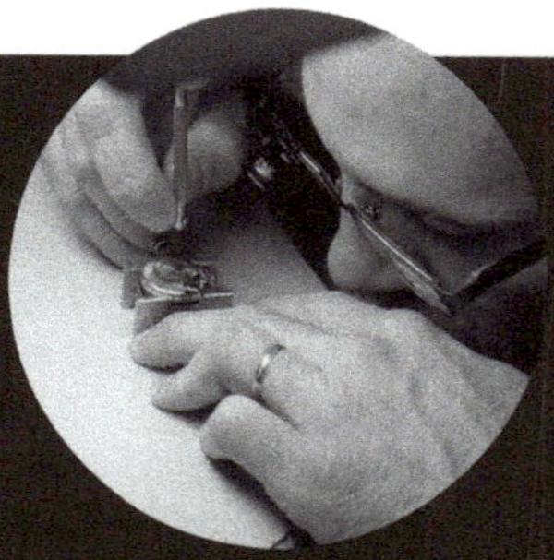

While many professionals make a living working on vintage watches, an elite group of approximately 140 in the USA, carry the prestigious CW21 certification. Awarded by the venerable American Watchmakers-Clockmakers Institute, a Certified Watchmaker 21st Century (CW21), is considered the highest and most respected technical standard a watchmaker can achieve.

This prestigious award was bestowed on Phillip Ridley in 2015, a second-generation master-watchmaker. Phillip is considered one of the best-kept secrets among elite vintage collectors. They jealously guard their relationship with him referring only their closest friends and associates, for fear of overwhelming him with work and losing their prized access to his services! This exclusive status has allowed him to work on some of the rarest and finest examples of the most prized vintage watches.

"No two vintage pieces are equal – meaning with each watch, there is a strategic game plan put together by the owner and myself to execute the repairs without the possibility of remorse in the future. A lot of this is done by educating the client and the trust they bestow upon me."

Phillip honed his skills under the mentorship of Frank Poye and the guidance of his father and has dedicated his career to sustaining the art of horology. He is most widely known for his comprehensive vintage movement restoration, and contemporary movement servicing. While he also has an outstanding track record working on dials, cases, and bracelets, his first love is vintage movements.

Like many of the new generation of up and coming watchmakers, Phillip has grown an extensive professional and social network, giving him access to an exclusive club of increasingly media-shy dealers and collectors. With these relationships he has been able to connect vintage buyers with sellers, ensuring that only bonafide, legitimate vintage Rolex watches in premium condition are changing hands.

"When it comes to vintage Rolex or anything designed well, preventive maintenance is critical to ensure the proper life span and retain its monetary value. Without doing so, only time will tell before the elements from the outside come within. I always inform clients to get a second opinion when purchasing their new vintage piece. There is usually a 72-hour window in which if anything is incorrect, or not as described, the buyer has the opportunity to return the watch."

Phillip can be reached via his website at ridleywatchmakers.com advice says you should service your watch every 5-10 years, depending on the era of the movement.

Even if correctly oiled and diligently but sparingly wound, parts will eventually wear out. Critical components were designed to be consumable and replaced during service intervals. These include barrel springs and pivot jewels. It is common and accepted by all but the most obsessed collectors, to use aftermarket non-Rolex consumables in antique and vintage movements.

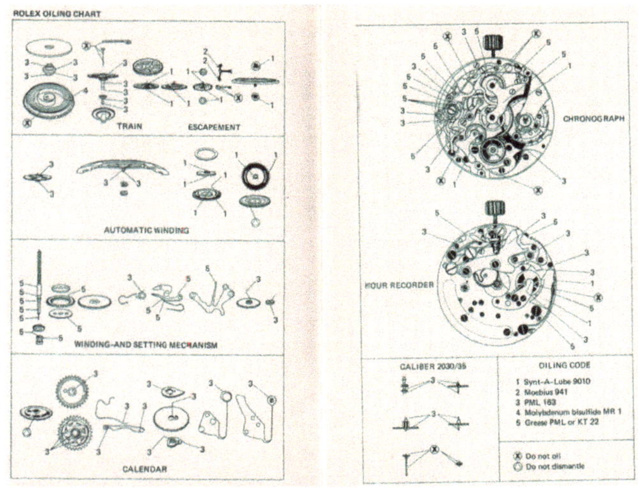

Rolex lubrication schedule, circa 1960

FINISHING

Antique and vintage Rolex are not generally known for refined movement finishing. The practice is usually associated with Haute Horlogerie, which embodies the ideal pursuit of design excellence and watchmaking as high art.

Rolex has always finished their movements, but to a lesser degree than other Swiss watchmakers. Their aim seemed to be to stamp their visual identity on ébauche movements and visually distinguish their pieces from those made by other watch companies rather than show off their finishing prowess. Except for Cellini, Rolex has never used see-through casebacks to showcase their movement finishing.

Movement finishing includes a variety of techniques applied to individual components once they've been stamped or milled out of base metal. All signs of machining or tool marks, such as burred edges, are removed by polishing. Components are then painstakingly decorated by hand.

Some types of finishing can have a practical application, but it's mostly used for aesthetics. However, electroplating can help prevent corrosion, and Geneva Stripes are sometimes said to trap dust away from moving parts. Heat treating steel screws turns them a deep royal blue color, but will also harden them. While there may be some minor engineering advantages, these techniques are just signs of the incredible care and attention watchmakers have given to the tiniest details.

Geneva Stripe is a traditional type of decoration consisting of regular parallel wave-like patterns applied to plates, bridges, and rotors. The stripes can be straight or circular but are always aligned perfectly across different parts. These most often appear on Rolex dress watches.

Vertical Geneva Stripes, Anglage bevels and screw sinks

Anglage refers to beveling or chamfering and is found on antique and vintage Rolex movements. It involves beveling edges of parts (in general to 45 degrees) like bridges and plates. The beveled edge can be meticulously polished, emphasizing the shape of the component and should be regular and consistent across all parts.

Perlage is another finish seen on antique and vintage Rolex. It is also known as circular-graining or stippling, and consists of covering a surface with a pattern of overlapping small circles using a rotating peg. This technique is seen on casebacks, but also movement base plates.

Perlage-style finishing on 18karat gold

Sinks are a concave chamfer around screw holes or seats for jewels. These chamfered rims of the drilled holes are often hand-polished.

Not all Rolex movements had the same degree of finish or type of finish, but all had some. It takes considerable experience to recognize Rolex finishing and know what's correct for a specific movement and period.

The broad description that follows highlights significant movements in references that are popular with collectors. The omission of significant calibers used in Ladies watches and Dress watches is not a judgment of their quality or significance.

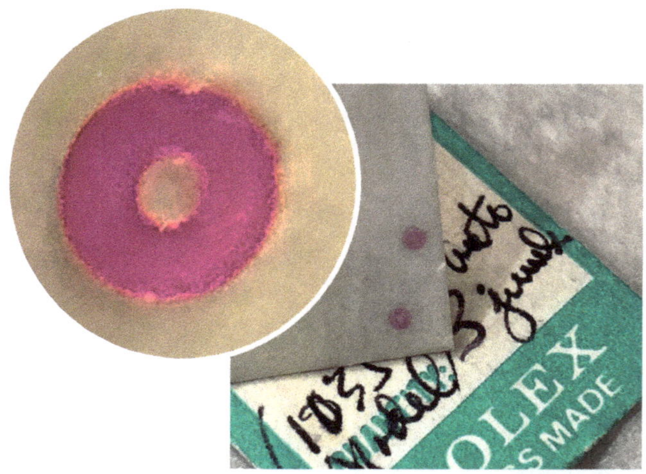

MOVEMENT JEWELS

The jewel count of a movement was once considered a relative indication of value and quality – more jewels implying higher quality and value. This was from a time when jewels were literally crafted from natural gems. With the advent of synthetic gemstones and industrial production techniques, this is no longer true. However, the myth persists.

Watch jewels are references to jewel bearings or jewel pivots. Invented in 1704 by Nicolas Fatio de Duillier, Peter Debaufre, and Jacob Debaufre, they were crafted from natural diamond, sapphire, ruby, or garnet and made by grinding the natural gem using diamond abrasive.

In 1902, Auguste Verneuil, invented a process to make synthetic sapphire and ruby (crystalline aluminum oxide, also known as corundum). This dramatically reduced their cost encouraging watchmakers to use them indiscriminately, even to the detriment of the watch. This allowed them to brag about their high jewel count. Today watch jewels are made with high-powered lasers, chemical etching, and ultrasonic milling.

During World War II, jewel bearings were one of the products restricted by the British War Production Board as critical to the war effort. The import and export of watch movements and parts were heavily restricted and taxed. Wilsdorf and Davis may have anticipated this in relocating their business to Switzerland at the outbreak of WWII.

The jewel itself is a small (tiny) concave shaped washer seated in a sunken pivot hole, in which an axle or pivot would spin. These jewel bearings offer low and predictable friction, good temperature stability, and the ability to operate without lubrication in corrosive environments.

A typical time-only watch has 17 jewels: two cap jewels, two pivot jewels, an impulse jewel for the balance wheel, two pivot jewels, two pallet jewels on the tips of the pallet fork, and two pivot jewels each for the escape, fourth, third, and center wheels. It is the two pallet fork jewels striking the steel teeth of the escape wheel that produce the familiar ticking sound of a mechanical watch.

In the vintage context, a 17 jewel movement could be timed or calibrated to higher precision than a 15 (or lesser) jewel movement, and by implication, was the higher quality, more valuable movement.

APPROXIMATE MOVEMENT TIMELINE

Rolex has never published the production history for their movements. However, collectors, enthusiasts, and watchmakers have been able to compile an approximate timeline using case serial numbers and records from other watch manufacturers that used similar ébauche. These are approximations of the year of introduction and should be used a rough guide only.

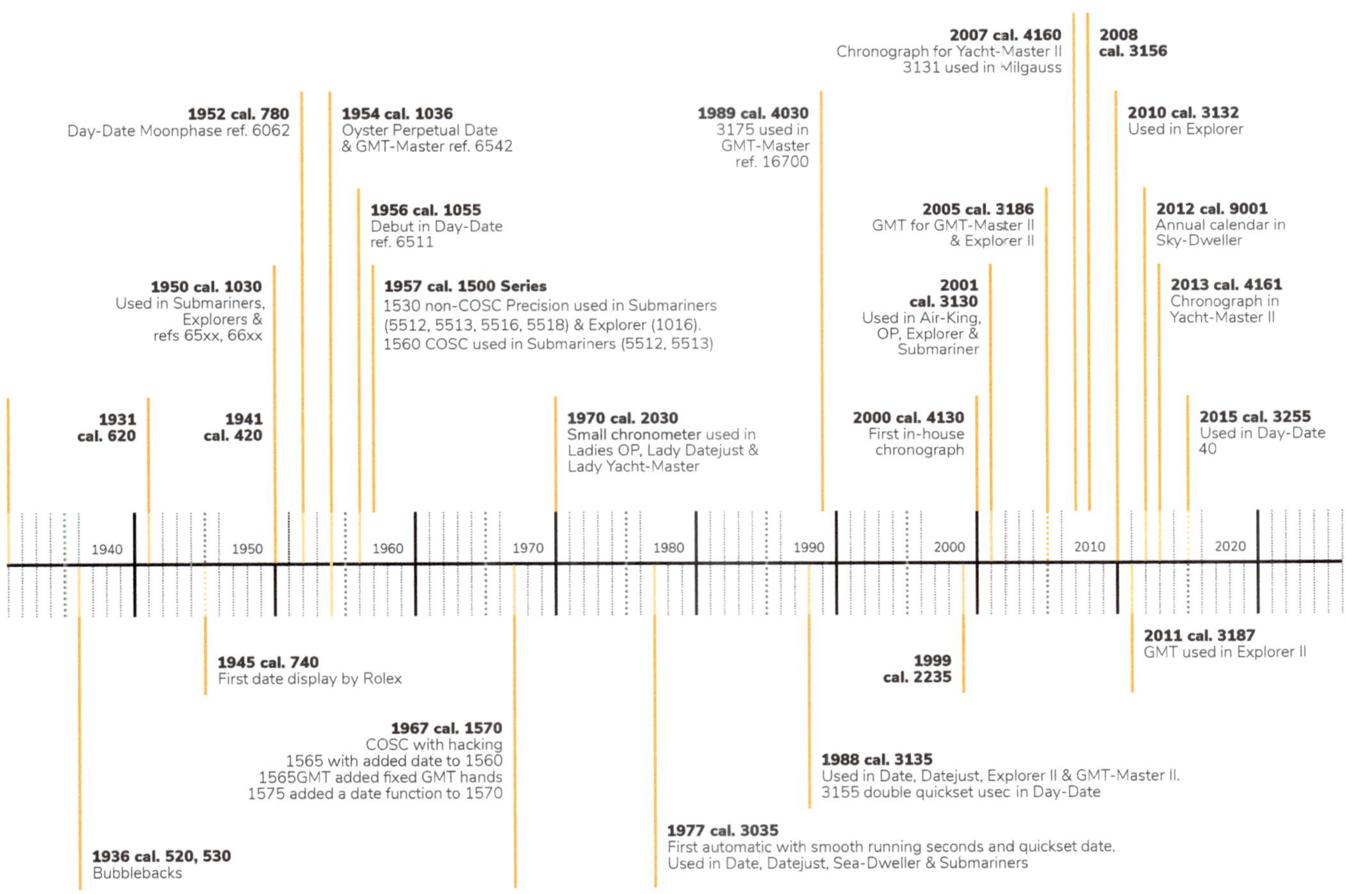

ANTIQUE MOVEMENTS

In 1905, Hans Wilsdorf placed the largest order for movements that Aegler had ever received. These were for Rebberg movements in two grades – 15 and 7 Jewel, all machine-finished. This order established a business partnership that would last a century and end in a business merger.

The movements Aegler supplied to Rolex in the early days (the antique period) fall into one of two categories – the Rebberg and the Hunter.

The name "Rebberg" was registered as a trademark in 1902 and was an homage to the Rebberg part of Bienne, home of the Aegler factory. Rebberg-class movements were used first and were available with both lever and cylinder escapements. The lever escapement is considered the more refined and collectible of the two movements.

Later, the Hunter movement replaced the aging Rebberg. Production and use of the Hunter would run almost twice as long as the Rebberg. Introduced in 1923 in 9½ ligne and 10½ ligne sizes, they had a polished rhodium finish.

Caliber 161 typical of Ladies cocktail watches from the 1950s & 60s

The first Hunter was a 15-jewel movement made in three grades: Prima, Extra Prima, and Ultra Prima. Later iterations with 16, 17, and 18 jewels were the first movements capable of achieving chronometer precision, and these watches bore the "Chronometer" designation on the dials.

VALJOUX

Valjoux (for Vallée de Joux, or "Joux Valley") is a historic Swiss manufacturer of mechanical watch movements. They are known for their mid- to high-end chronograph ebouche. Today, what remains of Valjoux assets and intellectual property belong to the Swatch Group (ETA).

The evolution and genealogy of their movements are complicated, and caliber numbering is convoluted. Manufacturers like Rolex often renamed (or renumbered) their Valjoux movements after customizing them for their own purposes.

The Rolex story of the Valjoux begins with the VJ23 and then its derivatives the VJ72 and 722 and 727.

Rolex first renamed the manual VJ72 as cal. 722-1, then 727. The VJ727 powered the Daytona from 1961 (ref. 6239) to 1988 (ref. 16520). It was eventually replaced by the automatic caliber 4030, derived from an enhanced Zenith 400.

So while Valjoux had been producing chronograph movements through the antique-period, Rolex began using them in earnest during the vintage period.

The table below describes Valjoux production beyond that used by Rolex. Many of these non-Rolex movements find their way into Rolex cases, either as complete donor movements, or spare parts.

VALJOUX CALIBERS

Caliber	Size (Ligne)	Rate (bph)	Period	Est. Prod. Volume	Description
13	13	18k			Mono-pusher, column-wheel, 30-minute counter
22	14	18k	1914–1966	129,010	Mono or Twin Pusher, column-wheel, 30-minute counter
23	12	18k	1916–1974	126,582 inc cal. 72c	Mono or Twin Pusher, column-wheel, 30-minute counter
69	10½	18k	1936–1966	3,964	Miniature chronograph, Twin Pusher, column-wheel, 30-minute counter
72	13	18k	1938–1974	261,558	Twin Pusher, column-wheel, 30-minute & 12-hour counters
72c	13	18k	1946–1974		Twin Pusher, column-wheel, 30-minute & 12-hour counters, simple full calendar
88	13	18k	1947–1974	13,218	Twin Pusher, column-wheel, 30-minute & 12-hour counters, simple full calendar and moon-phase indicator
230	13	18k	1954–1974	14,700	Annular or screw balance
232	13	18k	1969–1974	5,176	Caliber 23, date display with rapid switching
234	13	21.6k	1970–1974	16,458	Caliber 23, date display with rapid switching
235	13	21.6k	1971–1974	1,580	Caliber 23 with permanent return-to-zero function
236	13	21.6k	1969–1974	22,543	Caliber 23
237	13	21.6k	1971–1974	300	Caliber 23 with permanent return-to-zero function, date display with rapid switching
725	13	21.6k	1971–1974	1,400	Caliber 72 with permanent return-to-zero function
726	13	21.6k	1969–1974	47,379	Caliber 72
729	13	21.6k	1970–1974	1,300	Caliber 72 with 12-hour & 24-hour counters
730	13	21.6k	1971–1974	40,300	Caliber 72c

VALJOUX CALIBERS

VALJOUX 727

The column wheel chronograph of the Daytona. The caliber number 727 is engraved in the hub sink. Note the distinctively engraved bridge.

A generic Valjoux 23 in a Rolex ref. 2511 mid-size chronograph. It is a generic donor movement making this watch a Franken at best, and an outright counterfeit at worst.

VINTAGE MOVEMENTS

In 1935, Rolex was awarded patent 188,077 for the Superbalance. This patent leads to the three Prima grades being retired. Aegler manufactured the Superbalance Hunter movement in partnership with Rolex in 15 and 17 jewel configurations until 1969.

The differences between Superbalance movements were all in the quality of the jewels and pivots. The various grades had adjustment points in the form of small screws that could be tightened to alter the performance characteristics of moving parts. These adjustment screws are located on the main plate and seen easily if the caseback is removed. The simplest movements had two adjustment screws, while higher grade movements could have up to seven adjustments.

The automatic, self-winding Oyster Perpetual movement first appeared in 1931. It was created by fitting a semi-circular winding rotor to the Hunter 9 ¾ ligne. In 1944, the Oyster Perpetual mechanism was applied to the 10 ½ ligne Hunter and renamed the Rolex cal. 720, which ran until 1950.

CALIBER 1030 (1950–1957)

In 1950, the cal. 720 and Hunter series was retired and replaced by the entirely new base cal. 1030. There was a base cal. 1000, but few details are available and it didn't survive long. The cal. 1030 was a 25 jewel affair and represented a significant engineering leap forward.

This movement featured "Rolex Perpetual" on the bi-directional butterfly-style rotor. This movement was adjustable to 5 positions and was able to reach COSC levels of performance before the existence of the COSC testing authority. A date complication was added around 1952 and became cal. 1035.

The cal. 1030 (and 1035) are identifiable by this uniquely styled rotor. It has two angled cuts, which end in circles resembling a butterfly. The two cuts were designed to eliminate rotor shake, which occurred during excessive movement, causing the rotor to twist, which stressed the axel and made contact with the case. The cuts allowed the rotor to flex and absorb the motion. The 1030 series was very successful and provided the platform for the 1530 base caliber that replaced it in 1957.

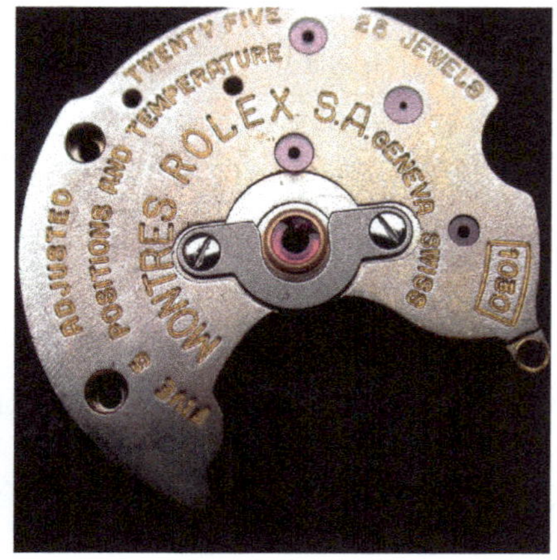

CALIBER 1030

Top: Movement in case | Auto-wind removed | Device bridge
Middle: Reverser wheels | Rear of butterfly rotor
Bottom: Movement in case | Disassembly

CALIBER 1200 (1954–1984)

The 1200 series were manual-winding movements based on the Hunter A720. Their reputation for being lesser movements of lower quality than the ref. 1030 is unfounded. They were merely different, less advanced designs. Seasoned watchmakers will attest to the accuracy these robust movements can attain – often within COSC standards (-4 /+6 secs per day) and even achieving Rolex's standards of -1/+5 secs per day.

The 1200 series received the Precision designation similar to the A720 Hunter Prima designations. The top 10% most precise movements received the Super Precision designation. The remaining 90% are Extra Precision and Precision grades.

Rolex continued to sell these manually wound Precision movements alongside their COSC certified automatic, Perpetual chronometers, like ref. 1035.

The 1200 series of movements were mechanically simple and very robust. They eventually received an upgrade in 1967 before being finally withdrawn in 1984.

Caliber 1210 (18,000 bps) 1954 to 1964

Caliber 1215 (18,000 bps) with Date 1954 to 1964

Caliber 1225 (26,600 bps) with Date 1967 to 1984

Aegler made these for Rolex in enormous numbers for use in 32mm and 34mm Oyster Precision and Oysterdate models. These relatively simple watches are inexpensive and very popular among collectors.

Two examples of the cal. 1210. Movement serial numbers N25063 (above), # 29053 (below).

CALIBER 1530 (1957–1963)

The cal. 1530 had a higher beat rate, better accuracy, and power reserve than cal. 1030 and 1035, which it replaced. It featured numerous incremental technical improvements. The series also received a date complication with cal. 1535.

A day of the week complication was later added to become the ref. 1555 (1959 to 1967) and used in the Day-Date President models. It later received an upgrade to cal. 1556 (1965 to 1978) along with 25 and 26 jewel versions.

The ref. 1530 has a flat semi-circular rotor with ring-shaped cuts and two red click wheels, making it easy to identify. It went through several upgrades before being retired in 1965.

The 1530 series (including the 1550 derivative) was hugely successful and ran for 20 years before being phased out across all models by 1977.

CALIBER 1520 (1963–1977)

Curiously, the ref. 1520 came after the ref. 1530 in 1963. It appeared to be an attempt to simplify and reduce the production costs of the ref. 1530. Despite being a high performer, it was never submitted for COSC certification but instead used in watches marked "Precision".

It later received a date function with the cal. 1525, and the two powered all of the non-chronometer watches until 1980. By then, all but the Air-King had gained COSC certification, and the 3000 series had arrived. The cal. 3000 was to become the standard issue for Rolex for many years.

CALIBER 1560 (1959–1965)

The ref. 1560 has come to be known as the start of the second generation of the 1500 series (the first generation being cal. 1530 and 1520 from 1963 to 1977). Its appearance and use overlapped with its predecessor, the ref. 1520.

It is a 26 jewel self-winding automatic movement with a date function added in the cal. 1565. A 24-hour hand was later added to make it the 1560GMT. Although there was no quick set mechanism, the system was able to create an instantaneous date change at midnight.

The first, no-date cal. 1560 movement was used in the Oyster Perpetual series beginning with ref. 1002 in the late 1950s. The movement's modest size (5.75mm height, 28.5mm diameter, 12.5 ligne) allowed it to fit comfortably in the 34mm Mid-Size and 36mm Full-Size cases of the era.

The Explorer ref. 1016 (1963 to 1989) initially launched with the cal. 1530, but was upgraded to the more accurate and chronometer-certified 1560. The Submariner 5512 followed a similar path.

Caliber 1560 Movement in case | Reverser wheels | Autowind removed

CALIBER 1570 (1965–1974)

Cal. 1570 marked the third generation of the 1500 series and introduced a hacking feature. It was now possible to stop the sweeping second hand by pulling out the crown and activating a hacking lever that interrupted the balance wheel. This innovation allowed for the accurate setting of the time and synchronizing with other timing instruments.

The ref. 1570 has become one of the most highly regarded Rolex movements and is a feature of some of the most desirable vintage references. A date function and GMT-hand was added and deployed in the Explorer II, Sea-Dweller 1665 and GMT-Master.

CALIBER 1570

Top: Movement in case | Reverser wheels | Dial removed.
Middle: Balance spring | Auto-wind mechanism removed
Bottom: Bridge engravings | Autowind cutouts | Reverser wheels

MODERN CLASSIC MOVEMENTS

CALIBER 3000 (1990–2001)

Cal. 3000 first appeared around 1977 and began replacing the 1500 series. It was a long-running process with both movements overlapping well into the 1990s. The cal. 3000 features in the Air-King 14000 series, the Submariner 14060, and Explorer no-date models. The COSC version, cal. 3130 became the upgrade path for models like the Submariner 14060M and Explorers 14000, 14010, 14060, and 14270.

CALIBER 3035 & 3135 (1977–PRESENT)

The 3035 was the first to be fitted with a quickset date wheel, making its debut in a Datejust. The cal. 3135 arrived in 1988 as a modest engineering evolution of the 3035. Modern classic watches with sapphire crystals used the caliber 3035 and 3135 throughout the 1990s.

The cal. 3135 has been used more broadly in more references than any other movement. It is highly finished compared to its predecessors and quite large at 28.5mm diameter and 6mm height. It also has a high frequency (28.8 Hz) and jewel count (31), making it consistently accurate.

Industry observers and commentators consider cal. 3135 one of the most commercially successful and highest-performing calibers produced by Rolex.

3135 SPECIAL NOTE

Cal. 3135 is a 31 jewel automatic movement and is 28.50 mm in diameter, and 6.00 mm height, with a frequency of 28.800 vph (vibrations per hour). The date complication has instantaneous rollover at midnight. The calendar bridge hidden under the dial has a perlage-style finish, and to reduce friction the calendar ring is mounted on 3 jewels. The balance wheel is made of glucydur.

Glucydur is the trade name for a beryllium bronze alloy. It is extremely hard, anti-magnetic, corrosion resistant, and has low thermal expansion properties, making it ideal for precision parts like balance wheels that must remain stable in a wide range of conditions.

On the balance there are two pairs of adjusting nuts known as the Microstella system. Rolex uses a Breguet overcoil in the 3135 for better isochronism (consistent periodic intervals, or oscillations). The Breguet overcoil, combined with the Perpetual winding mechanism ensures time consistency irrespective of mainspring tension. So the watch continues to tell accurate time even as the main spring is almost entirely unwound.

The surfaces of bridges are rhodium plated and beveled. Screws are finished to a high degree, featuring polished edges, which are seated in chamfered slots.

Almost all parts are made by Rolex with the exception of the hairspring, which is made by Nivarox.

Glucydur was introduced around the same time as Nivarox and both have very similar properties. The Nivarox patent belongs to the Swatch Group, who continues to enjoy patent royalties.

MOVEMENT REFERENCES

Base	Cal	Period	Jewels	Type	Rate	Shape	Engraving	Date	Description
0	59	-	17	Manual	18000	Round		None	Fontmelon used in Oyster Watch Co and Canadian Junior Sport and Tudor
0	2766	1977–1977	0						
0	2230	-	31	Automatic	28800				
23	23	1960–1960	17	Manual Chronograph	18000	Round		None	Valjoux twin register chronograph
23	72	-	17	Manual	18000	Round		None	
23	72A	1964–1964	17	Manual Chronograph	18000	Round	Balance Wheel	None	Shock Protecting
23	72B	1968–1968	17	Manual Chronograph	18000	Round	Balance Wheel	None	Shock Protecting
23	72C	1962–1962	17	Manual Chronograph	18000	Round	Balance Wheel	None	Shock Protecting
60	65	-	0	Manual					Tudor
80	80	-	0	Manual					Tudor
80	89	-	0	Manual					Tudor
90	90	1961–1961	17	Manual		Rectangular		None	
90	92	1961–1961	17	Manual					Tudor
90	94	1961–1961	17	Manual					Tudor
90	99	1961–1961	17	Manual					Tudor
100	100	1945–1945	17	Manual		Rectangular		None	Super Balance
100	120	-	17	Manual					Tudor
100	123	-	17	Manual					Tudor
100	144	-	17	Manual					Tudor
100	147	-	17	Manual					Tudor
100	131	-	17	Manual					Tudor
100	150	1950–1950	17	Manual				None	
100	160	1947–1947	17	Manual		Rectangular		None	Super Balance
100	161	1950–1950	17	Manual	18000	Rectangular	Plate	None	Super Balance
100	170	1952–1952	17	Manual	18000	Rectangular		None	Princess
100	180	1950–1950	17	Manual	18000	Elongated		None	Super Balance
100	190	1952–1952	17	Manual	18000	Rectangular	Balance Wheel	None	
100	191	1953–1953	17	Manual	21000	Rectangular	Balance Wheel	None	
100	192	1956–1956	17	Manual	21600	Rectangular	Balance Wheel	None	
100	193	1961–1961	17	Manual	21600	Rectangular		None	
200	200	1938–1938	17	Manual		Oval		None	Extra Prima
200	210	1945–1945	17	Manual		Oval		None	Prima
200	219	-	0	Manual					Tudor

Base	Cal	Period	Jewels	Type	Rate	Shape	Engraving	Date	Description
200	234	1977–1977	0	Manual Chronograph				Date	Tudor
200	250	1945–1945	15	Manual		Oval		None	Extra Prima
200	270	1950–1950	17	Manual	18000			None	
200	276	-	0	Manual					Tudor
200	280	1954–1954	17	Manual	21600		Balance Wheel		
200	281	1958–1958	17	Manual	21600		Balance Wheel		
200	282	1963–1963	17	Manual	21600		Balance Wheel		Shock Protecting
200	290	-	15	Manual	21600		Balance Wheel		Tudor non-shock
200	290	-	17	Manual			Balance Wheel		Tudor Shock resisting
300	300	1938–1938	18	Manual Chronometer		Rectangular		None	
300	310	1938–1938	18	Manual Chronometer		Rectangular		None	
300	320	1938–1938	0	Manual				None	Tudor
300	340	-	17	Manual	18000		Balance Wheel	None	Tudor Shock resisting
300	342	-	21	Manual	18000		Plate	None	Tudor Shock resisting
300	343	-	17	Manual	21600	Rectangular	Balance Wheel	None	Tudor Shock resisting
300	343	-	21	Manual	21600	Rectangular	Balance Wheel	None	Tudor Shock resisting
300	350	1938–1938	18	Manual	18000	Rectangular	Balance Wheel	None	Extra Prima
300	350	1938–1938	17	Manual	18000	Rectangular	Balance Wheel	None	Tudor Shock Resisting
300	360	1953–1953	18	Manual Chronometer	18000	Rectangular	Balance Wheel	None	Super Balance
300	374	-	15	Manual	18000		Plate	None	Tudor non-shock
300	374	-	17	Manual	18000		Balance Wheel	None	Tudor Shock resisting
300	374	-	21	Manual	18000		Plate	None	Tudor Shock resisting
300	375	-	17	Manual			Plate	Date	Tudor Shock resisting
300	380	-	17	Manual					Tudor
300	390	-	17	Automatic				None	Tudor
300	395	-	17	Automatic				Date	Tudor
400	400	1953–1953	17	Manual				None	Super Balance
400	420	1952–1952	18	Automatic					
400	425	1952–1952	17	Automatic			Plate	Date	Tudor
400	434	1952–1952	0						Tudor
500	500	1953–1953	17	Manual				None	Super Balance, Precision
500	510	1953–1953	17	Manual				None	Super Balance, Precision
500	520	1948–1948	18	Automatic					
500	530	1948–1948	17	Automatic					
600	600	1951–1951	17	Manual				None	Precision
600	605	-	0						Tudor

Base	Cal	Period	Jewels	Type	Rate	Shape	Engraving	Date	Description
600	611	-	0						Tudor
600	618	1974–1974	0						Tudor
600	620	1951–1951	18	Automatic Chronometer				None	Super Balance
600	622	1968–1968	0						Tudor
600	624	1974–1974	0						Tudor
600	630	1951–1951	18	Automatic				None	
600	635	1954–1954	18	Automatic				None	Shock Protecting
600	640	1954–1954	18	Automatic				Moonphase	
600	645	1954–1954	18	Automatic				None	
600	650	1977–1977	17	Manual	18000	Ultra-slim	Train Wheel Bridge	None	
600	651	1992–1992	18	Manual	21600	Ultra-slim	Train Wheel Bridge	None	
700	700	1951–1951	18	Manual Chronometer				None	Super Balance
700	710	1951–1951	17	Manual				None	Super Balance, Precision
700	720	1952–1952	18	Automatic Chronometer				None	Super Balance
700	722	1968–1968	0		18000	Round			Shock Protecting
700	727	1984–1984	17	Manual Chronograph	18000	Round		None	Shock Protecting
700	740	1955–1955	18	Automatic Chronometer		Round		Date	
700	730	1952–1952	18	Automatic Chronometer	21600	Round		None	
700	745	1955–1955	0	Automatic		Round		Date	
700	750	1992–1992	20	Manual	21000		Barrel Bridge	None	Pocket Watch
700	760	1956–1956	0	Manual				Date	
700	765	1955–1955	18	Manual				Date	
700	775	1955–1955	18	Manual				None	
700	780	-	0	Manual				Moonphase	
700	722-1	1969–1969	17	Manual Chronograph				None	Shock Protecting
800	800	1960–1960	17	Manual Chronometer				None	Ultra Prima, Pocket Watch, Shock Protecting
800	832	1960–1960	17						Tudor
800	850	1960–1960	17	Manual Chronometer				None	Pocket Watch, Shock Protecting
1000	1000	1955–1955	18	Manual	18000		Balance Wheel	None	Shock Protecting
1000	1001	1955–1955	17	Manual	18000		Balance Wheel		Tudor, Schock Resisting
1000	1007	1955–1955	17	Manual					Tudor
1000	1012	1955–1955	17	Manual					Tudor

Base	Cal	Period	Jewels	Type	Rate	Shape	Engraving	Date	Description
1000	1016	1985–1985	0	Automatic	18000		Self-winding Bridge		
1000	1030	1957–1957	17	Automatic	18000	Round	Self-winding Bridge	None	Shock Protecting
1000	1030	1957–1957	25	Automatic	18000	Round	Self-winding Bridge	None	Shock Protecting
1000	1035	1957–1957	25	Automatic	18000	Round	Self-winding Bridge	Date	Shock Protecting
1000	1036	1957–1957	25	Automatic	18000	Round	Self-winding Bridge	Date	Shock Protecting
1000	1040	1957–1957	26	Automatic	18000	Round	Self-winding Bridge	None	Shock Protecting
1000	1040	1957–1957	25	Automatic	18000	Round	Self-winding Bridge	None	Shock Protecting
1000	1051	1956–1956	0						
1000	1055	1956–1956	25	Automatic	18000		Self-winding Bridge	Day Date	Shock Protecting
1000	1057	1956–1956	0						Tudor
1000	1061	1956–1956	0						Tudor
1000	1065	1959–1959	25	Automatic	18000	Round	Self-winding Bridge	Date	Shock Protecting
1000	1066	1957–1957	25	Automatic	18000	Round	Self-winding Bridge	Date	Shock Protecting
1000	1080	1956–1956	25	Automatic	18000	Round	Self-winding Bridge	None	Anti-magnetic, Shock Protecting
1000	1080	1956–1956	0						Tudor
1000	1036 GMT	1957–1957	25	Automatic	18000	Round	Self-winding Bridge	Date	Shock Protecting
1000	1055B	1956–1956	25	Automatic	18000	Round	Self-winding Bridge	Day Date	Shock Protecting
1000	1065 GMT	1957–1957	25	Automatic	18000	Round	Self-winding Bridge	Date	Shock Protecting
1000	1065M	1959–1959	25	Automatic	18000	Round	Self-winding Bridge	None	Shock Protecting
1000	1066 GMT	1957–1957	25	Automatic	18000	Round	Self-winding Bridge	Date	Shock Protecting
1000	1066M	1957–1957	25	Automatic	18000	Round	Self-winding Bridge	None	Anti-magnetic, Shock Protecting
1100	1100	1957–1957	17	Manual	19800	Round	Balance Wheel	None	Shock Protecting
1100	1120	1956–1956	27	Automatic	19800	Round	Self-winding Bridge	None	Shock Protecting
1100	1130	1967–1967	26	Automatic	19800	Round	Self-winding Bridge	None	Shock Protecting
1100	1135	1967–1967	26	Automatic	19800	Round	Self-winding Bridge	Date	Shock Protecting
1100	1151	1967–1967	0						Tudor
1100	1156	1967–1967	17						Tudor
1100	1156	1967–1967	21						Tudor

Base	Cal	Period	Jewels	Type	Rate	Shape	Engraving	Date	Description
1100	1160	1968–1968	26	Automatic	19800	Round	Self-winding Bridge	None	Shock Protecting
1100	1161	1970–1970	26	Automatic	19800	Round	Self-winding Bridge	None	Shock Protecting
1100	1165	1968–1968	26	Automatic	19800	Round	Self-winding Bridge	None	Shock Protecting
1100	1166	1970–1970	26	Automatic	19800	Round	Self-winding Bridge	None	Shock Protecting
1100	1170	-	21	Manual	18000	Round	Balance Wheel	None	Shock Resisting
1100	1173	-	17	Manual	18000	Round	Balance Wheel	None	Shock Resisting
1100	1182	-	0						Tudor
1100	1187	-	0						Tudor
1200	1200	1964–1964	17	Manual	18000	Round	Balance Wheel	None	Shock Protecting
1200	1210	1967–1967	17	Manual	18000	Round	Balance Wheel	None	Shock Protecting
1200	1215	1967–1967	17	Manual	18000	Round	Balance Wheel	None	Shock Protecting
1200	1216	1967–1967	17	Manual	18000	Round	Balance Wheel	None	Oysterdate, Shock Protecting
1200	1220	1984–1984	17	Manual	21600	Round	Balance Wheel	None	Shock Protecting
1200	1225	1984–1984	17	Manual	21600	Round	Balance Wheel	None	Oysterdate, Shock Protecting
1200	1250	-	0						Tudor
1200	1260	-	0						Tudor
1200	1272	-	0						Tudor
1200	1280	-	0						Tudor
1200	1240	-	0	Manual					Shock Protecting
1300	1300	1967–1967	17	Manual	18000	Rectangular	Balance Wheel	None	Shock Protecting
1300	1301	1967–1967	0						Tudor
1300	1310	1967–1967	18	Manual	18000	Rectangular	Balance Wheel	None	Shock Protecting
1300	1315	1967–1967	17	Manual	18000	Rectangular	Balance Wheel	Date	Shock Protecting
1300	1320	1967–1967	0						Tudor
1400	1400	1984–1984	18	Manual	21600		Balance Wheel	None	Shock Protecting
1400	1401	1966–1966	18	Automatic	21600		Balance Wheel	None	Shock Protecting
1400	1410	-	0	Automatic	21600			None	Tudor, Auto Princess, Shock Resisting
1400	1411	-	0	Automatic	21600			None	Tudor, Auto Prince Junior, Shock Resisting
1400	1412	-	0	Automatic	21600			Date	Tudor, Princess Date, Shok Resisting
1400	1413	-	0	Automatic	21600			Date	Tudor, Prince Junior, Shock Resisting
1400	1432	-	0						Tudor
1400	1475	-	0						Tudor
1500	1506	-	0						Tudor

Base	Cal	Period	Jewels	Type	Rate	Shape	Engraving	Date	Description
1500	1520	1984–1984	26	Automatic	19800		Self-winding Bridge	None	Shock Protecting (EU Version)
1500	1525	1984–1984	26	Automatic	19800		Self-winding Bridge	Date	Shock Protecting
1500	1530	1964–1964	17	Automatic	18000		Self-winding Bridge	None	Shock Protecting (US Version)
1500	1530	1964–1964	25	Automatic	18000		Self-winding Bridge	None	Shock Protecting (EU Version)
1500	1535	1967–1967	26	Automatic	18000		Self-winding Bridge	Date	
1500	1536	1977–1977	26	Automatic				None	
1500	1555	1967–1967	26	Automatic	18000		Self-winding Bridge	Day Date	Shock Protecting
1500	1556	1978–1978	26	Automatic	19800		Self-winding Bridge	Day Date	Shock Protecting
1500	1560	1967–1967	26	Automatic	18000		Self-winding Bridge	None	Shock Protecting
1500	1565	1967–1967	26	Automatic	18000		Self-winding Bridge	Date	Shock Protecting
1500	1566	1967–1967	17	Automatic	18000		Plate	None	Shock Resisting
1500	1566	1967–1967	25	Automatic	18000		Hub sinking	None	Shock Resisting
1500	1570	1979–1979	26	Automatic Chronometer	19800		Self-winding Bridge	None	Shock Protecting
1500	1575	1979–1979	26	Automatic	19800		Self-winding Bridge	Date	Shock Protecting
1500	1580	1979–1979	26	Automatic	19800		Self-winding Bridge	None	Shock Protecting
1500	1565 GMT	1967–1967	25	Automatic	18000	Round	Self-winding Bridge	Date	Shock Protecting
1500	1575 GMT	1979–1979	26	Automatic	19800	Round	Self-winding Bridge	Date	Shock Protecting
1600	1600	1984–1984	19	Manual	19800		Balance Wheel	None	Shock Protecting
1600	1601	1979–1979	20	Manual	19800		Balance Wheel	None	Shock Protecting
1600	1602	2006–2006	20	Manual	21600		Train Wheel Bridge	None	Shock Protecting
1800	1800	1974–1974	17	Manual	21600		Balance Wheel	None	Shock Protecting
1800	1895	1977–1977	21	Automatic	21600		Balance Wheel	Day Date	Shock Resisting
2000	2030	1979–1979	28	Automatic	28000		Self-winding Bridge	None	Shock Protecting
2000	2035	1979–1979	28	Automatic	28000		Self-winding Bridge	Date	Shock Protecting
2100	2130	1992–1992	29	Automatic	28000		Self-winding Bridge	None	Shock Protecting
2100	2135	1992–1992	29	Automatic	28000		Self-winding Bridge	Date	Shock Protecting
2300	2235	2010–2010	0						Tudor

Base	Cal	Period	Jewels	Type	Rate	Shape	Engraving	Date	Description
2400	2402	-	0						Tudor
2400	2403	-	0						Tudor
2400	2411	1977–1977	17	Manual	21600		Balance Wheel		Tudor, Shock Resisting
2400	2412	1977–1977	17	Manual	21600		Plate		Tudor, Shock Resisting
2400	2414	1977–1977	17	Manual	18000		Plate		Tudor, Shock Resisting
2400	2422	-	0						Tudor
2400	2423	-	0						Tudor
2400	2424	-	17	Manual	21600		Balance Wheel		Tudor, Shock Resisting
2400	2446	-	0	Automatic	18000		Balance Wheel		Tudor, Auto-Princess, Shock Resisting
2400	2447	-	0	Automatic	18000		Balance Wheel	Date	Tudor, Auto-Princess, Shock Resisting
2400	2461	-	25	Automatic			Plate		Tudor, Shock Resisting
2400	2462	-	25	Automatic			Plate	Date	Tudor, Shock Resisting
2400	2483	-	0	Automatic	18000		Balance Wheel		Tudor, Shock Resisting
2400	2484	-	0	Automatic	18000		Balance Wheel	Date	Tudor, Shock Resisting
2500	2510	-	17	Manual	18000		Balance Wheel	None	Tudor, Shock Resisting
2500	2512	1977–1977	17	Manual	21600		Balance Wheel	None	Tudor, Shock Resisting
2500	2546	-	17	Manual	21600		Balance Wheel	None	Tudor, Shock Resisting
2500	2554	-	0	Automatic	21600		Balance Wheel	None	Tudor, Shock Resisting
2500	2555	-	0	Automatic	21600		Balance Wheel	Date	Tudor, Shock Resisting
2600	2650	1977–1977	0	Automatic					Tudor, Shock Resisting
2600	2651	1977–1977	0	Automatic					Tudor, Shock Resisting
2600	2671	1977–1977	21	Automatic			Balance Wheel	Date	Tudor, Shock Resisting
2700	2750	1977–1977	17	Manual	21600		Balance Wheel		Tudor, Shock Resisting
2700	2772	-	0	Automatic					Tudor
2700	2776	-	25	Automatic	21600		Balance Wheel	Date	Tudor, Shock Resisting
2700	2784	1977–1977	0	Automatic					Tudor
3000	3000	1992–1992	27	Automatic		Round			OP, Air King, Explorer, Submariner
3000	3035	1990–1990	27	Automatic	28800	Round	Self-winding Bridge	Date	Shock Protecting
3000	3055	1990–1990	27	Automatic	28800	Round	Self-winding Bridge	Day Date	Shock Protecting
3000	3075	1992–1992	27	Automatic	28800	Round	Self-winding Bridge	Date	12/24 Hr hand
3000	3085	2004–2004	27	Automatic	28800	Round	Self-winding Bridge	Date	12/24 Hr hand
3100	3130	-	31	Automatic	28800	Round		None	OP, Air King, Explorer, Submariner
3100	3135	2001–2001	31	Automatic	28800	Round	Self-winding Bridge	Quick Set	Datejust, Submariner Date, Sea Dweller, Yacht Master
3100	3155	2001–2001	31	Automatic	28800	Round	Self-winding Bridge	Day Date	Day Date

Base	Cal	Period	Jewels	Type	Rate	Shape	Engraving	Date	Description
3100	3175	1992–1992	31	Automatic	28800	Round		Quick Set	GMT
3100	3185	1992 - 1992	31	Automatic	28800	Round	Self-winding Bridge	Quick Set	GMT II
3400	3475	1977 - 1977	0			Round			Tudor
4000	4030	1992 - 1992	31	Automatic	28800		Chrono Bridge	None	Daytona, Shock Protecting
4000	4130	-	44	Automatic	28800			None	Daytona Chronograph
5000	5035	1992 - 1992	11	Quartz	32768		Bridge	Quick Set	Datejust
5000	5055	1992 - 1992	11	Quartz	32768		Bridge	Quick Set	Day-Date
6000	6620	1992 - 1992	8	Quartz			Bridge	None	Ultra-slim
6000	6620	1992 - 1992	8	Quartz					Tudor
6000	6630	1977 - 1977	0						Tudor
7000	7000	1977 - 1977	21	Manual	18000	Round	Balance Wheel	None	Tudor, Shock Resisting
7000	7016	1977 - 1977	21	Manual		Round	Balance Wheel	Date	Tudor, Shock Resisting
7000	7733	-	0	Manual		Round		Date	Tudor Chornograph, date at 6:00
7000	7734	-	0	Manual		Round		Date	Tudor Chornograph, date at 6:00
7000	7750	1977 - 1977	17	Automatic	28800	Round	Plate	Date	Tudor Chronograph
7000	7750	1977 - 1977	25	Automatic	28800	Round	Plate	Date	Tudor Chronograph

MONTRES
ROLEX S.A.
GENEVA SWISS

1953-1954

Anglage | äng-lä-gə |

noun ***art, metalwork, French***

- the process of making a chamfer.
- a 45 degree angled edge of a workpiece.

Chapter 12

ACCESSORIES

The term full-set describes the accessories that accompany a new watch when first sold. These include all the original packaging, booklets, sales receipts, hang tags, and warranty papers. These artifacts add value to the implied provenance and history of the watch, and some collectors will obsess over these items as much as the watches themselves.

Counterfeit accessories are as common as the watches they accompany. A counterfeiter may add genuine accessories to throw a buyer off the scent of a fake watch, or a seller may cobble together accessories to goose the price of a sub-par, authentic watch.

As a general rule, the older the watch, the less likely it is to have the original accessories and antique watches are unlikely to have their original boxes and warranty papers at all. These older watches are lucky to have survived a century of wear (and war), and offers of original documents and packaging are improbable at best.

Vintage era watches may have surviving accessories from the 1950s. Collectors prize these for their nostalgia rather than any intrinsic value. Any price premium varies with the popularity of the watch model as well as the condition of both the watch and accessories. For example, old Submariner boxes and papers are more attractive than Datejust equivalents.

Modern Classics are quite likely to include their accessories. Many loose items are circulating on eBay, which can quickly be cobbled together to make a set. The unwary buyer may miss the cues of an incorrect and hurriedly assembled collection of mismatched accessories.

There is a vibrant market for accessories and collectors enjoy the challenge of assembling them correctly. Verifying that a set of accessories are period correct, authentic, and original to a watch can be a challenging and technical task. Some love this challenge while others choose to ignore accessories all together.

The specific type and combination of accessories vary with the era and model of watch. All had inner and outer boxes, warranty papers and booklets of various sorts. COSC watches had hang tags, and dive watches had other unusual items like anchors and bracelet tools.

Buyers are discouraged from placing too much emphasis on accessories. After all, they can't be worn and will likely spend their time in the darkness of a drawer or safe. However, they are worth acquiring if you plan to resell the watch. If you plan to own the watch long term, any price premium is entirely personal and subjective.

"Two things are infinite: the universe and human stupidity; and I'm not sure about the universe."

Albert Einstein

BOXES

There are a wide variety of boxes that have been used over the years. All were made to look and feel luxurious with velvet and faux leather. Service boxes, pouches, and booklets also exist but are more functional and less luxurious. These too, varied by region and era.

During the vintage era, boxes were made by subcontractors and paired with the watch at the point of sale by the jeweler or dealer so it is not unusual to see inconsistencies, particularly between different territories. Generally, men's watches were sold in green boxes and ladies' in red. Special boxes were used for the top of the line, jeweled models.

Box packaging should include both a luxurious inner box and an outer cardboard box, printed in various different styles. The style of box is collection specific, and print markings and identification codes changed over the years. All inner boxes feature a product number on their base. To date, there is no consensus amongst collectors as to what they indicate.

Distressed or corroded boxes can be restored and refinished. Several craftsmen (and women) offer their services through social media. The quality of their work is high, and they can return a box to a condition that's almost as good as new. Restoration includes replacing faux leather exterior coverings and reupholstering interior cushions, padding, and linings.

Counterfeit boxes are common and can be easily purchased online. They can be difficult to identify and distinguish from authentic ones unless you have handled lots of genuine examples. The best way to guard against them is to know exactly what style is period correct for the reference you're pursuing. Another tip is to become familiar with the aged smell of vintage wood lined boxes. This distinctive smell can be difficult to counterfeit.

The Bufkor Box circa 1970s

One such independent manufacturer of packaging is Bufkor Inc. founded in 1895. Now a part of the SigmaQ group of companies who specialize in luxury and jewelry packaging.

This three-piece plastic box is enclosed in a green and gold cardboard outer box. It is most commonly associated with sports professional models sold in the N. American markets.

BOOKLETS

There are generally three booklets included with a watch when sold new – a model specific brochure, an Oyster Bracelet brochure, and a generic translation of the guarantee and warranty commitment.

The model brochures have a date of printing on the second to last page. This date of printing should be within one or two years of the production date of the watch to be considered correct. Booklets are easy to counterfeit, so look for appropriate aging and fading and be sure you know exactly what style is period correct for the watch it accompanies.

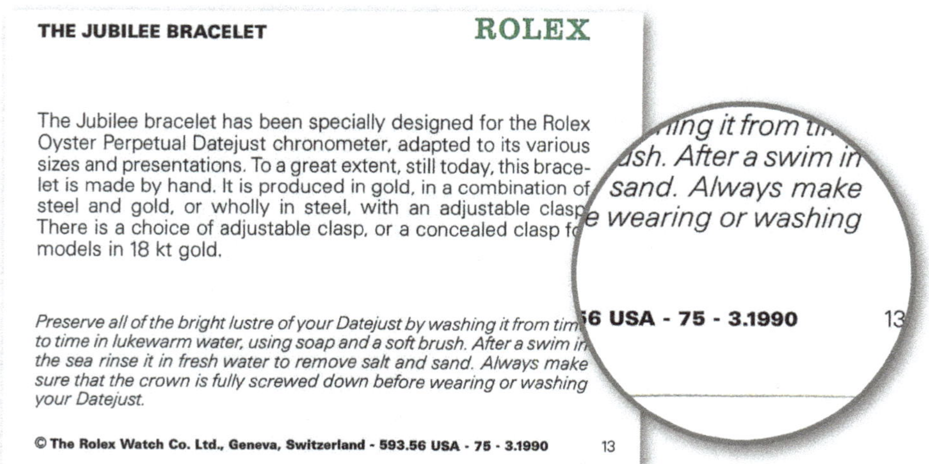
THE JUBILEE BRACELET ROLEX

The Jubilee bracelet has been specially designed for the Rolex Oyster Perpetual Datejust chronometer, adapted to its various sizes and presentations. To a great extent, still today, this bracelet is made by hand. It is produced in gold, in a combination of steel and gold, or wholly in steel, with an adjustable clasp. There is a choice of adjustable clasp, or a concealed clasp for models in 18 kt gold.

Preserve all of the bright lustre of your Datejust by washing it from time to time in lukewarm water, using soap and a soft brush. After a swim in the sea rinse it in fresh water to remove salt and sand. Always make sure that the crown is fully screwed down before wearing or washing your Datejust.

© The Rolex Watch Co. Ltd., Geneva, Switzerland - 593.56 USA - 75 - 3.1990 13

HANG TAGS

There are two types of hang tags a buyer can expect to encounter – a chronometer tag and a green "Oyster Swimpruf" tag.

The chronometer tag first appeared in the mid-1940s as a small, dark red disk, resembling a wax seal attached to a decorative string. The tag indicates that the watch movement has passed COSC certification and performs to chronometer standards. These tags replaced the paper COSC testing certificates that were phased out.

The red COSC tag went through several iterations. The first version was an all-red disc with a swirling, marbled appearance. The text was embossed, and the decorative string (or twine) was burgundy and gold. This style lasted into the early 1960s.

The second version had the same embossed text but overprinted in gold. The dark maroon colored plastic had a less obvious marble-effect, and the string was green and gold. Around 1967, the string color again changed back to the red and gold of the first version. This second variant lasted until about 1970.

In the 1970s, the coronet changed to the distinctive frog-foot style, and the embossed word, "Genève", became "Geneva-Bienne".

In the 1980s the coronet style changed again to a wide style. In 1989, around the R-serial range, the red tag acquired a hologram depicting a single coronet. In 2000, this hologram changed to include multiple coronets. This multi-crown hologram persisted through the transition to the green chronometer tag in 2015.

In July 2015 the chronometer tag became green, marking the change of the Rolex warranty from two years to five. All modern models will have a green chronometer tag and a multi-crown hologram. The green tags can have a hologram with either large or small multi-crowns.

ANCHORS

Around 1969, Rolex redesigned their packaging and began including an anchor trinket with their dive watches. This item is a small, keyring-like miniature anchor, bearing the depth rating of the watch with meters on one side and feet on the other. They were included with Submariners and Sea-Dwellers until around F-serials, in 2005.

Although no longer produced, anchors have become coveted despite not having any particular purpose. Anchors typically appear in a silver steel color, but a gold colored version accompanied the solid gold Submariner.

Fake anchors are known to exist, presumably made to accompany fake Submariners and Sea-Dwellers.

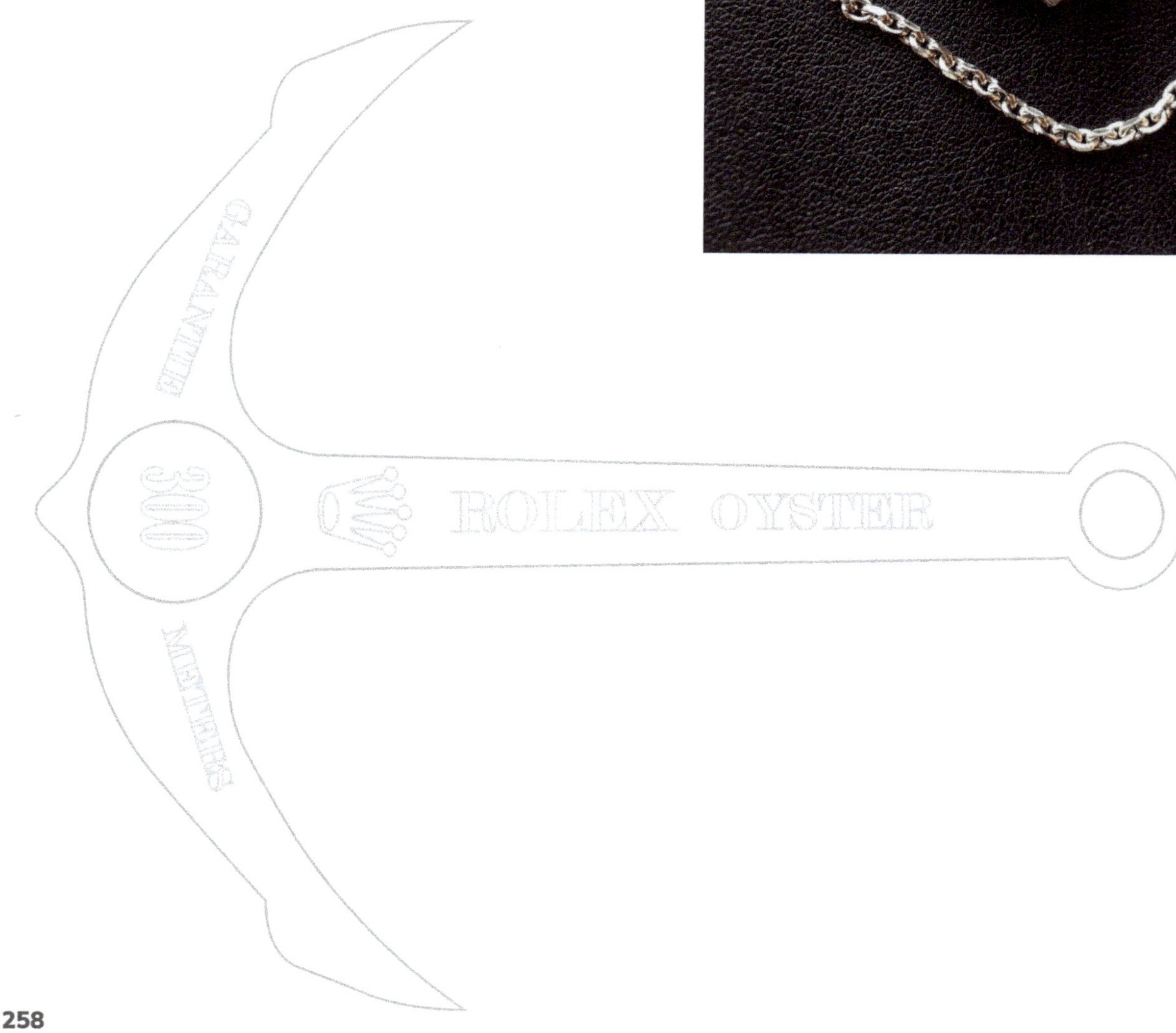

WARRANTY PAPERS

There is a misconception that Rolex warranty or guarantee papers ensure (with absolute certainty) the authenticity of a vintage watch. This assumption is not valid and old warranty papers, sales, and service receipts are not guarantees of a watch's authenticity. However, they can contain useful data that may help validate the watch and reduce the degree of uncertainty.

Warranty and guarantee documents usually bear the details of the jeweler, many of whom have long ceased to exist. It would not be unreasonable to try and contact the jeweler to see if they have any record of having sold the watch, but as the antique period pre-dates the official Rolex dealer network, this may not be practical or possible. So while the papers are interesting, they're of little value for determining authenticity.

Vintage warranty papers appear in several styles, languages, and sizes. Most have visible watermarks embedded in them and these should be visible if the certificate is held up to a window or bright light.

It's not uncommon for old papers to have the watch details handwritten with a ballpoint pen. These are easily washed. Soaking the paper in acetone (nail polish remover) will remove the ink and the solvent will dry without warping or marking the paper. Washed documents can be found on eBay and used for all manner of shady purposes. Typewriter ink is much harder – but not impossible – to remove. As a result, collectors prefer punched-style papers over handwritten. However, vintage certificate punching machines are still in circulation from long-defunct dealers, and these can be used to punch serial numbers onto washed or forged papers.

Warranty paper punched numbers are 4 x 6 holes. Counterfeit papers are sometimes punched with 5 x 7 or 5 x 9 holes.

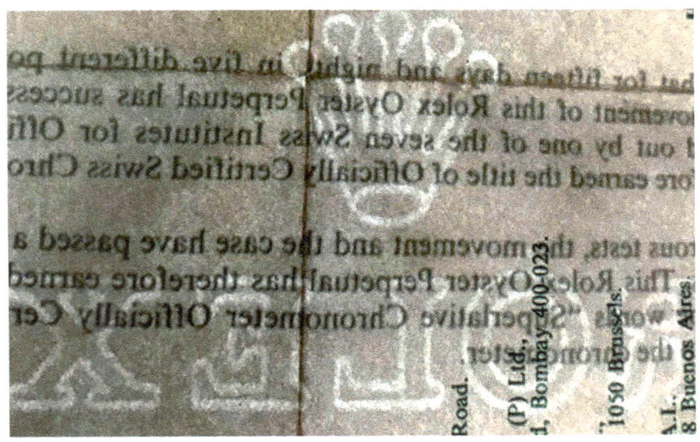

Warranty paper watermark

Vintage Rolex warranty paper punch issued to Authorized Dealers.

Original documents are desirable and helpful, but the absence of any papers is preferable to suspicious, mismatched, or fake papers. You should be aware that it is possible to convincingly engrave a counterfeit Oyster case with serial and reference numbers that match a genuine warranty paper. This unsavory practice has been observed at the high end of the vintage market with top-dollar, highly desirable early Daytonas.

Watches sold in the USA had warranty papers featuring some codes that are unique to the US market. A 14-digit "Style No" is often printed (or typed) above the "Serial No" e.g., "R16520A50B7839". This Style No. can be parsed into six fields – (L) NNNNN(N) X NX (L XXNN).

(L) BRAND: R = Rolex or T = Tudor

NNNNN(N) WATCH REFERENCE: Five or six digits, e.g. 16528, 116520 etc.

X CASE MATERIAL:

A = Stainless steel (from French acier)

3 = Gold and steel

4 = Stainless steel and white gold

8 = Yellow gold

9 = White gold

NX DIAL STYLE:

0U = Black Mother of Pearl with diamonds

0W = Black Mother of Pearl Roman

10 = Silver

18 = Silver Tapestry

20 = Champagne

28 = Champagne Tapestry

30 = Black

3H = Onyx Arabic with 2 diamonds

50 = White

51 = White Roman

52 = Ivory Pyramide Roman

57 = White Arabic

83 = Rose Roman

8X = Onyx Serti

9K = White Mother Pearl Arabic

9R = Mother Pearl Roman

L BRACELET

B = Bracelet or S = Strap. For integrated bracelet (like Oysterquartz) this code is missing.

XXNN BRACELET TYPE

Four digits or two characters and two digits e.g. 7839, 7876, 9315, etc.

In the example (R16520A50B7839) this would translate to:

R	Rolex (or T for Tudor)
16520	Reference number for a Daytona with a Zenith Calibre
A	Stainless steel
50	White dial
B	Bracelet
7839	Bracelet ref. 78390

In addition to the Style No., Rolex USA guarantee papers prior to 2000 had a further code on the back printed in red. These five digits indicate the date Rolex USA shipped the watch to the dealer.

Letter	**Key**
R	1
O	2
L	3
E	4
X	5
W	6
A	7
T	8
C	9
H	0

So a code like "C OWCE" would be "9 2694", which translates to September 26 '94.

Ephemera | ə'fem(ə)rə |

noun ***historical***

- something of no lasting significance.
- paper items that were originally meant to be discarded after use but have since become collectibles.
- objects that when produced, were not intended to last a long time.

ROLEX
7 WORLS'S RECORDS
GOLD MEDAL
GENEVA - SUISSE

✠ 114948

1926

Chapter 13
EPIGRAPH

And so you have it – as much data as any reasonable Rolex enthusiast can digest in one sitting.

You should now have a sense for the variety of watches circulating in the wild, and the many pitfalls in the path between you and them. Along the way, you will encounter crooks and conmen, spivs, and posers – all seeking to take your money and your pride. Fakes, replicas, and counterfeits are increasingly common in the antique, vintage pre-owned markets.

- 15-30% of internet searches on watches involve people looking for replicas. (Source: Forbes, 2013) People are increasingly seeking the faux-vintage aesthetic.
- In 2017, Switzerland exported 25 million watches. China exported 663 million. While there are some well-known Chinese brands in this number, none have the demand or cachet of Rolex.
- The Swiss Customs Service claims that there are some 30 to 40 million counterfeit watches entering circulation each year (2005).
- Swiss Customs estimates that 40% of fake watches come from China (2006). Irrespective of the accuracy of this estimate, large numbers of replica watches are being made or assembled elsewhere, including the US and EU.

According to a 2012 Federation of Swiss Watches estimate, counterfeit Swiss watch sales generated over a S1 billion in sales per year.

There are a lot of people buying fake watches and there's no shortage of supply. Counterfeiters are willing to risk jail and financial penalties to profit from this demand. Replica watches are illegal to buy and sell and are typically of poor quality. They can be found in the fake-districts of every major city around the world and, increasingly, online.

In the pre-owned market, the phenomena of the Super Fake continues to grow. Fakes are not a harmless imitation or an homage, but a product of organized crime.

If you are even remotely inclined to buy a fake vintage Rolex, then this book and the watches discussed here are not for you. Don't feel bad if you're missing the vintage watch gene or the disposable income – you're just not ready for vintage Rolex. Instead, spend your money on a more affordable and genuine vintage Longines, Seiko, Zenith, or any other number of increasingly collectible brands. You'll be much happier.

Fortune favors the brave, and your life is only getting shorter, so find a genuine vintage Rolex you like, make sure you can afford it, and then wear the heck out of it. Admire it, show it off, enjoy it, share it, and love its well-worn and unoriginal peculiarities.

If you've gotten to this point in the guide, I'd like to thank you for buying it and taking the time to read it. I hope it's helpful in your journey. If you use it to buy a watch, I would love to hear from you. If you have feedback, corrections, or even more data, please get in touch. And if you have time, please leave a 5-star review on Amazon, irrespective of where you bought the book from.

You can find me on Instagram: **@morning_tundra** or through old-fashioned email: **morningtundra@gmail.com**

Bonne Chance!

"Obsessed is a word the lay use to describe the dedicated"

Anonymous

Collector | kə'lek-tər |

noun

- a person who collects things of a specific type, professionally or as a hoby.
- a member of an order or ***fraternity***.

www.ingramcontent.com/pod-product-compliance
Lightning Source LLC
LaVergne TN
LVHW070831220926
848486LV00030B/271